Carel Faber
Eduard Looijenga (Eds.)

Moduli of Curves and Abelian Varieties

Aspects of Mathematics

Edited by Klas Diederich

*A Publication of the Max-Planck-Institut für Mathematik, Bonn

Carel Faber
Eduard Looijenga
(Eds.)

Moduli of Curves and Abelian Varieties

The Dutch Intercity Seminar on Moduli

Prof. Dr. *Eduard Looijenga*
Dept. of Mathematics
University of Utrecht
PO Box 80. 010
NL-3508 TA Utrecht

Dr. *Carel Faber*
Dept. of Mathematics
Oklahoma State University
Stillwater, OK 74078-1058
USA

and

Dept. of Mathematics
Royal Institute of Technology
S-100 44 Stockholm

Prof. Dr. *Klas Diederich* (Series Editor)
Dept. of Mathematics
University of Wuppertal
D-42119 Wuppertal

Vieweg is a subsidiary company of Bertelsmann Professional Information.

http://www.vieweg.de

Cover design: Ulrike Weigel, www.CorporateDesignGroup.de

ISBN 978-3-322-90174-3   ISBN 978-3-322-90172-9 (eBook)
DOI 10.1007/978-3-322-90172-9

Preface

The present volume, with contributions of R. Dijkgraaf, C. Faber, G. van der Geer, R. Hain, E. Looijenga, and F. Oort, originates from the Dutch Intercity Seminar on Moduli (year 1995–96). Some of the articles here were discussed, in preliminary form, in the seminar; others are completely new. Two introductory papers, on moduli of abelian varieties and on moduli of curves, accompany the articles.

Topics include a stratification of a moduli space of abelian varieties in positive characteristic, and the calculation of the classes of the strata, tautological classes for moduli of abelian varieties as well as for moduli of curves, correspondences between moduli spaces of curves, locally symmetric families of curves and jacobians, and the role of symmetric product spaces in quantum field theory, string theory and matrix theory.

This Intercity Seminar is part of the long term project "Algebraic curves and Riemann surfaces: geometry, arithmetic and applications", sponsored by the Netherlands Organization for Scientific Research (NWO), that has been running since 1994. Its ancestry can be traced back to joint activities in the seventies (if not earlier), which as of 1980 had evolved into active biweekly research seminars. These have been a focal point of Dutch algebraic geometry and singularity theory since.

We are grateful to NWO for its support for the project. C.F. thanks the Max-Planck-Institut für Mathematik, Bonn, for support during the final stages of the preparation of this volume.

Bonn/Utrecht, May 1999

Carel Faber
Eduard Looijenga

Contents

Fields, Strings, Matrices and Symmetric Products   151

Robbert Dijkgraaf

Moduli of Abelian Varieties: A Short Introduction and Survey

Gerard van der Geer and Frans Oort

Abstract

We give a short introduction to the moduli of abelian varieties as well as a survey of recent research and open questions.

1 Introduction

The term moduli was introduced in 1857 by Riemann to describe the essential parameters of Riemann surfaces of given genus. The use of the term has since then been extended to a name for the essential parameters of algebraic varieties (of a given type). Moduli spaces are playing an increasingly important role in mathematics and since a number of years also in mathematical physics. Not only can we learn about families of algebraic varieties of a certain type by studying their moduli, but these moduli spaces are themselves most interesting mathematical objects with intriguing arithmetic properties. Often moduli spaces can be understood much better than arbitrary varieties because their invariants and properties have an intrinsic meaning, which makes them much more amenable to treatment; certain deep conjectures about algebraic varieties can be proved for certain moduli spaces, but still defy verification in general.

The moduli spaces most studied so far are the moduli of elliptic curves. That they possess fascinating arithmetic properties was recognized in the late 19th century and by now they have become a central object of study in number theory.

An elliptic curve is a (regular, complete, absolutely irreducible) curve of genus 1 together with a marked point. It can be given as a regular cubic curve in the projective plane. The crucial point is that these curves carry the structure of a group variety, obtained by identifying the points of the curve with the group of divisor classes of degree 0 on the curve.

The concept 'elliptic curve' admits two natural generalizations, depending on whether we stick to dimension 1 or want to retain the notion of a group variety:

- a curve of genus g,
- an abelian variety of dimension g.

These generalizations lead to two types of moduli spaces: the moduli spaces $\mathcal{M}_g$ of curves of genus g and the (various) moduli spaces of (polarized) abelian varieties. These spaces have been the focus of intense study in the last decades. As a result we learned a lot about them, though it seems fair to say that our understanding of them is still very limited and that many basic and important questions are still open.

Although Riemann introduced the concept of moduli already in the middle of the 19th century and although the moduli of curves were studied by Italian geometers around the turn of the century, the concept was first put on a firm foundation around 1960 by Grothendieck, who introduced the notion of representability of a functor. This made it possible to give a precise meaning to 'moduli'. The existence of moduli spaces in this sense for curves and polarized abelian varieties was shown in the sixties by Mumford in his book [45].

2   Elliptic Curves

Let us illustrate these concepts for elliptic curves. An elliptic curve E over a field is a complete, geometrically irreducible regular curve of genus 1 with a marked point. If P is the marked point then the linear system defined by $3P$ defines an embedding of the curve E into the projective plane and the image is given as a smooth plane cubic curve; with the image of P equal to the point at infinity $(0:1:0)$ we may write the equation for $E - \{P\}$ in affine form as

$$y^2 + a_1 xy + a_3 y = x^3 + a_2 x^2 + a_4 x + a_6,$$

where the discriminant (a polynomial expression in the coefficients a_i) does not vanish. Conversely, any such regular plane cubic curve with the point P defines an elliptic curve. For the theory of elliptic curves we refer to [64]. Two such curves over an algebraically closed field k are isomorphic if and only if their j-invariants (a certain rational expression in the a_i) are the same: $E \cong E' \iff j(E) = j(E')$. So the isomorphism classes of such curves over k are in $1-1$ correspondence with the points of the affine line $\mathbb{A}^1$, the j-line. However, ideally one would like to have much more, namely a universal family of elliptic curves $\mathcal{X} \to \mathbb{A}^1$ such that for any (good) family of elliptic curves $X \to S$ there is a morphism $\nu : S \to \mathbb{A}^1$ such that $X = \nu^*(\mathcal{X})$.

Such a 'universal' family does not exist, but only by a narrow margin. The trouble is caused by the existence of automorphisms of our objects, elliptic curves. If we rule out these automorphisms by considering elliptic curves plus some extra structure (which rigidifies our objects) then these moduli spaces do exist: the functor F : Schemes $\to$ Sets which associates to a scheme S the set of isomorphism classes of elliptic curves (with enough extra structure) over S, is representable. This means that there exists a scheme $\mathcal{M}$ and a functorial isomorphism $F(S) =$ Mor$(S,\mathcal{M})$ for all S.

An example is given by the moduli spaces of elliptic curves with a level n structure, i.e. (assuming that n is invertible on our base scheme) an identification $E[n] \cong (\mathbb{Z}/n\mathbb{Z})^2$ of the kernel $E[n]$ of multiplication $\times n : E \to E$. The level n-structure for $n \geq 3$ rules out automorphisms of the elliptic curve and the corresponding moduli functors are representable. But this additional structure is just an artificial gadget, so we want to get rid of it.

By forgetting the extra structure the scheme $\mathcal{M}$ is replaced by a quotient space $M = \mathcal{M}/G$ with G a finite group. The existence of the universal family is then lost, unless one works with orbifolds. But the space M still possesses two important properties:

i) over an algebraically closed field k the points $M(k)$ are in $1 - 1$ correspondence with the isomorphism classes of elliptic curves;

ii) any transformation of functors $F \to$ Hom$(\cdot,M')$ factors uniquely through the transformation $F \to$ Hom$(\cdot,M)$.

The second condition means that the space we get is as close as possible to what we want. Such a space is called a *coarse moduli space*. However, if one is willing to work with 'stacks' or 'orbifolds', then the coarse moduli spaces are replaced by objects carrying a universal family.

The moduli spaces of elliptic curves over the field of complex numbers were studied in detail by Klein, Weber and others in the 19th century. As Riemann surfaces, these moduli spaces can be viewed as quotient spaces of the upper half plane of the complex plane $\mathbb{C}$ by a subgroup of finite index in SL$(2,\mathbb{Z})$. Their study was continued by Hecke, Eichler and Shimura, Igusa and many others. Eichler and Shimura also set the first steps in the study of characteristic p properties of these spaces using the work of Hasse and Deuring for elliptic curves in characteristic p. For a modern introduction we refer to [33].

The existence of these moduli spaces over $\mathbb{Z}$ is a very strong fact with many important applications. It makes it possible to study these spaces in all characteristics and also the reduction of characteristic 0 to characteristic p. One of the first instances where this was fruitful is the so-called Eichler-Shimura congruence relation which made it possible to express the zeta-functions of modular curves in terms of modular forms, cf. Shimura's book [62].

In studying elliptic curves in characteristic $p > 0$ one encounters totally new phenomena. For example in characteristic p the multiplication by p map $\times p : E \to E$ is inseparable. Since the map is of degree p^2 the degree of inseparability is p or p^2. If it is p the curve is called *ordinary*; if it is p^2 then the curve has no points of order p and the curve is called *supersingular*. (These curves are not singular, but their name refers to the fact that these elliptic curves have large endomorphism rings, larger than those with what classically are called the "singular j-values".) Such curves can be considered as in some sense quite degenerate curves. Hasse gave an equation for their j-invariant and Deuring gave a formula for their number. Deuring's formula says

$$\sum_{E/\cong} \frac{1}{\#\mathrm{Aut}(E)} = \frac{(p-1)}{24},$$

where the sum is over the supersingular elliptic curves over k (an algebraically closed field of characteristic p). This beautiful formula has clearly an 'orbifold' or 'stacky' flavour as the term $1/\#\mathrm{Aut}(E)$ shows and it suggests that when we view the moduli space just as the j-line we are losing information.

The special properties that one encounters in characteristic p are not only interesting in their own right, but also have applications in characteristic 0, as we shall see later.

A spectacular result on modular curves was obtained by Mazur, who determined the $\mathbb{Q}$-rational points on the modular curves $X_0(N)$. We refer to his paper [39]. Note that this is an example where we can determine a deep diophantine property of these curves, whereas for arbitrary curves this seems impossible.

3   Abelian Varieties

An abelian variety is a complete variety carrying the structure of an algebraic group. Such varieties are then automatically projective and the groups abelian. The easiest examples are the elliptic curves we dealt with before, complex tori which are algebraic, and last but not least the Jacobians of algebraic curves. For the theory of abelian varieties we refer to [46], [35]. But in order to describe their parameters in a meaningful way one must add further structure, as is known from the classification theory of varieties. This extra structure is given by the class of an ample divisor on an abelian variety. First we introduce the dual abelian variety.

For an abelian variety X (and more general for an abelian scheme: a "family of abelian varieties") one can define its dual abelian variety. In the algebraic context, this is $X^t := \mathrm{Pic}^0(X)$, the connected component of the Picard variety. In the complex analytic context, for a complex torus $T = X(\mathbb{C})$ we know that

$$X(\mathbb{C}) \cong t_X/\Lambda,$$

the quotient of the tangent space t_X by the exponential map $\exp : t_X \to X(\mathbb{C})$ with kernel a lattice $\Lambda_X \subset t_X$. The dual $T^t = X^t(\mathbb{C})$ is given as quotient of the complex anti-linear dual of t_X by the dual lattice, cf. [35]. We know that X^t is an abelian variety with $\dim(X) = \dim(X^t)$. Moreover, we know that the dual of the dual $(X^t)^t$ is canonically isomorphic to X itself.

A divisor D on X defines a morphism

$$\varphi_D : X \to X^t, \quad \text{by} \quad x \mapsto [D_x - D] \in \text{Pic}^0(X).$$

Here D_x is the translate of the divisor D over x. It is known that for an effective divisor D this is an isogeny (a surjective homomorphism with finite kernel) if and only if D is ample.

Definition: An isogeny $\lambda : X \to X^t$ is called a *polarization* if and only if there exists an effective divisor D on $X \otimes k$ (where k is an algebraically closed field) such that $\lambda_k = \varphi_D$. We say that $\lambda : X \to X^t$ is a *principal* polarization if it is a polarization which is an isomorphism.

Note that if $\lambda : X \to X^t$ is a polarization then it is *symmetric* in the following sense:

$$(\lambda : X \to X^t) = (\lambda : X \to X^t)^t = (\lambda^t : X = X^{tt} \to X^t).$$

An elliptic curve has a unique principal polarization (given by the divisor consisting of one point on the curve). An abelian variety in general does not admit a principal polarization (examples are easy to give) and some abelian varieties admit more than one principal polarization. Every abelian variety admits a polarization, and an abelian variety defined over an algebraically closed field is isogenous to an abelian variety which admits a principal polarization.

In the following we shall stick to the case of principally polarized abelian varieties of dimension g. We shall often denote an effective divisor giving rise to the polarization by Θ.

The basic result about the moduli spaces of principally polarized abelian varieties is the following.

Theorem 3.1. *(Mumford) For a given $g \in \mathbb{Z}_{\geq 0}$ the coarse moduli scheme $\mathcal{A}_g \to \text{Spec}(\mathbb{Z})$ of principally polarized abelian varieties of dimension g exists.*

See [45], 7.3, Th. 7.10.

Again, if we are willing to work with stacks then $\mathcal{A}_g$ carries a universal family $\mathcal{X}_g$. The (relative) dimension of $\mathcal{A}_g$ is $g(g+1)/2$. The tangent space at a point $[X]$ can be identified with $\text{Sym}^2(T_X)$ with T_X the tangent space to X at $0 \in X$.

Classically one had the following construction for the complex moduli of principally polarized abelian varieties. A complex abelian variety X is given by a lattice:

$$0 \to \Lambda_X \longrightarrow t_X \cong \mathbb{C}^g \xrightarrow{\text{exp}} X(\mathbb{C}) \to 0,$$

and a polarization is given by a hermitian form on t_X. If the polarization is principal, then the form is perfect. A choice of a basis of the lattice defines a period matrix, a complex symmetric matrix whose imaginary part is positive definite, i.e. an element of the Siegel upper half space:

$$\mathfrak{h}_g = \{M \in \text{Mat}(g \times g, \mathbb{C}) : \quad {}^t M = M, \text{Im}(M) > 0\}.$$

In view of the choice of a basis, the symplectic group $\Gamma = \mathrm{Sp}(2g,\mathbb{Z})$ operates in a natural way on $\mathfrak{h}_g$, and we have a representation as an arithmetic quotient of a bounded symmetric domain:

$$\mathfrak{h}_g \longrightarrow \Gamma \backslash \mathfrak{h}_g \cong \mathcal{A}_g(\mathbb{C}).$$

Since the action is not free (e.g. the element -1 acts trivially which reflects the existence of the automorphism -1_X), one should consider this quotient as an orbifold.

Since abelian varieties can degenerate, the moduli space $\mathcal{A}_g$ is not compact or complete. For example, a plane cubic curve can degenerate into a cubic with one ordinary double point. Classically, the moduli spaces of elliptic curves $\Gamma(n) \backslash \mathfrak{h}_1$, where $\Gamma(n) = \ker(\mathrm{SL}(2,\mathbb{Z}) \to \mathrm{SL}(2,\mathbb{Z}/n\mathbb{Z}))$ is the group associated to the level n structure described above, were compactified by adding the 'cusps' which correspond to such degenerate cubics. Satake constructed a compactification of $\mathcal{A}_g(\mathbb{C})$ in 1956, cf. [60]. His compactification is a normal analytic space $\mathcal{A}_g^*(\mathbb{C})$ which set-theoretically is the disjoint union

$$\mathcal{A}_g^*(\mathbb{C}) = \mathcal{A}_g(\mathbb{C}) \sqcup \mathcal{A}_{g-1}^*(\mathbb{C}) = \sqcup_{0 \leq j \leq g} \mathcal{A}_j(\mathbb{C}).$$

However, this space is very singular along the "boundary" $\mathcal{A}_g^*(\mathbb{C}) - \mathcal{A}_g(\mathbb{C})$. If an abelian variety degenerates, then the degenerate variety still might contain an abelian variety part (of lower dimension) and the boundary parametrizes only these lower dimensional abelian parts. The fact that in doing this so much of the structure is forgotten is reflected in the fact that $\mathcal{A}_g^*(\mathbb{C})$ is very singular. Baily and Borel generalized in [3] Satake's construction to a compactification of bounded hermitian symmetric domains and showed that these compactifications can be embedded in projective space (using modular forms).

Using the tools created by Mumford c.s., Faltings constructed in 1985 a smooth compactification of the moduli space $\mathcal{A}_g$ over $\mathbb{Z}$, (smooth $/\mathbb{Z}$ in the sense of stacks) cf. [20]. He also obtained the construction of the Satake compactification over $\mathbb{Z}$. The smooth compactifications thus constructed carry families of semi-abelian varieties. But they do not represent a functor. Compactifications that do represent a functor are constructed recently by Alexeev, cf. [1]; however, although his functors are quite natural, his spaces contain irreducible components that are much larger than the closure of $\mathcal{A}_g$.

For the modular curves parametrizing elliptic curves the main geometric invariants, like the genus, were determined long ago. Except for finitely many, these modular curves are of genus ≥ 2. This fact was generalized by Freitag, Tai and Mumford to the moduli spaces $\mathcal{A}_g$ (see [47]):

Theorem 3.2. *(Freitag, Tai, Mumford.) The moduli space $\mathcal{A}_g \otimes \mathbb{C}$ is of general type for $g \geq 7$.*

We know that the moduli space $\mathcal{A}_g$ is rational for $g = 1,2,3$ (for $g = 3$ we know this only since a few years by Katsylo, cf. [32]) and unirational for $g = 4,5$ by Clemens and Donagi. The Kodaira dimension is unknown for $g = 6$.

The moduli space $\mathcal{A}_g$ carries an interesting natural bundle, namely the bundle $\mathbb{E}$ defined as the cotangent bundle along the zero section $s : \mathcal{A}_g \to \mathcal{X}_g$. It is a rank g bundle. (Here we either pretend that the variety $\mathcal{A}_g$ possesses a universal family or else work with stacks.) This bundle is also called the 'Hodge bundle' since it can be identified with $\pi_*(\Omega^1_{\mathcal{X}_g/\mathcal{A}_g})$. The Chern classes $c_i(\mathbb{E})$ of $\mathbb{E}$ are denoted λ_i for $i = 1, \ldots, g$.

The determinant bundle $\det(\mathbb{E})$ is a classically well-known bundle, namely the bundle of modular forms. Over $\mathbb{C}$ the sections of $\det(\mathbb{E})^k$ can be described as holomorphic functions $f(\tau)$ on the Siegel upper half plane satisfying the transformation behaviour

$$f((a\tau + b)(c\tau + d)^{-1}) = \det(c\tau + d)^k f(\tau) \quad \text{for some } k$$

under elements

$$\begin{pmatrix} a & b \\ c & d \end{pmatrix} \in \mathrm{Sp}(2g,\mathbb{Z})$$

(plus some growth condition if $g = 1$). The integer k is called the weight of the modular form f. It is known that $\det(\mathbb{E})$ is an ample line bundle on the Satake compactification $\mathcal{A}_g^*$. As stated above, this was proved by Baily and Borel for the complex-analytic case, and the proof was extended to the general case by Moret-Bailly [44]. More precisely, the modular forms of an appropriate large weight define a morphism of a toroidal compactification $\widetilde{\mathcal{A}}_g$ to projective space whose image is the Satake compactification.

If we forget the quotient singularities, which are due to the fixed points of elements of finite order of $\mathrm{Sp}(2g,\mathbb{Z})/\pm 1$, the canonical bundle of a toroidal compactification $\widetilde{\mathcal{A}}_g$ is described as

$$K = \det(\mathbb{E})^{\otimes(g+1)} \otimes O(D),$$

where D is the divisor which compactifies $\mathcal{A}_g$, i.e. $D = \widetilde{\mathcal{A}}_g - \mathcal{A}_g$. What the theorem of Freitag, Tai and Mumford says is that for large g the ample line bundle $\det(\mathbb{E})$ is the dominant factor.

Although $\det(\mathbb{E})$ is ample, the bundle $\mathbb{E}$ is not (for $g \geq 2$). Nevertheless, its Chern classes possess certain positivity properties, e.g. they can be represented by effective cycles in characteristic p, see [24]. Using the Grothendieck-Riemann-Roch theorem one can prove the basic relation

$$(1 + \lambda_1 + \lambda_2 + \ldots + \lambda_g)(1 - \lambda_1 + \lambda_2 - \ldots + (-1)^g \lambda_g) = 1$$

between the Chern classes λ_i in the Chow ring $\mathrm{CH}_{\mathbb{Q}}$ of $\mathcal{A}_g$, cf. [24]. This extends the proof by Mumford of that relation for the cohomology classes λ_i, see [50]. Indeed, the Hodge bundle $\mathbb{E}$ fits into an exact sequence

$$0 \to \mathbb{E} \to \mathbb{H} \to \mathbb{E}^\vee \to 0,$$

where $\mathbb{H}$ is the bundle on $\mathcal{A}_g$ whose fibres are the de Rham cohomology groups $H^1_{\mathrm{dR}}(X)$ of an abelian variety. This bundle $\mathbb{H}$ carries an integrable connection which implies the relation in cohomology.

From these results it follows that the minimum codimension of a complete subvariety of $\mathcal{A}_g$ is g:

Theorem 3.3. *The dimension of a complete subvariety of $\mathcal{A}_g \otimes k$ with k a field is $\leq g(g-1)/2$.*

It can be shown that there exists complete subvarieties of codimension g in characteristic p: indeed, the subvariety of $\mathcal{A}_g$ whose points describe the abelian varieties that do not have points of order p has codimension exactly g in $\mathcal{A}_g \otimes \mathbb{F}_p$, cf. [34], [51]. Abelian varieties with no points of order p in characteristic p do not degenerate [52]. Essentially this is because a degeneration would contain a toric part (a multiplicative group), and these have geometric points of order p.

But these varieties can be degenerate in another sense, namely, they can be products of lower-dimensional principally polarized abelian varieties. A more precise analysis (cf. [19]) of the classes of product loci yields the following result:

Theorem 3.4. *A complete subvariety of $\mathcal{A}_g$ that does not contain points corresponding to products of principally polarized abelian varieties has codimension at least $2g - 1$ in $\mathcal{A}_g$.*

For $\mathcal{A}_g \otimes \mathbb{C}$ one has the following conjecture.

Conjecture 3.5. *(Oort) Let $g \geq 3$ and suppose that $W \subset \mathcal{A}_g \otimes \mathbb{C}$ is a complete subvariety. Then*
$$\dim(W) < g(g-1)/2.$$

By intersecting successively with hyperplanes one can construct complete subvarieties of dimension $g-1$. Indeed, $\mathcal{A}_g^*$ is a projective variety, the codimension of the boundary $\mathcal{A}_g^* - \mathcal{A}_g$ is g, so by intersecting suitable hyperplanes $g(g-1)/2+1$ times we can avoid the boundary. More explicitly, to obtain a complete subvariety of dimension $g - 1$ we can take the zero locus of an Eisenstein series on a Hilbert modular subvariety of $\mathcal{A}_g$ associated to a totally real field of degree g. It is an interesting question what the maximum dimension is of a complete subvariety which is the image in $\mathcal{A}_g$ of an arithmetic quotient of a bounded symmetric domain.

Using the expression for the canonical class and the proportionality theorem of Hirzebruch and Mumford's extension of it (cf. [49]) one can calculate the Chern numbers of (toroidal smooth compactifications of) $\mathcal{A}_g$. For example, the Euler number of $\mathcal{A}_g$ can be expressed (according to Siegel and Harder) as a product of values of the Riemann zeta function

$$\prod_{k=1}^{g} \zeta(1 - 2k).$$

The cohomology of such a (toroidal or Satake) compactification can be expressed in terms of Siegel modular forms. However, our understanding and knowledge of these forms is still so limited that it does not help much at this moment to understand the cohomology. Also the Chow rings are unknown except for small values of g. Mumford calculated the Chow ring of the (standard) compactification of $\mathcal{A}_2$ in [50] and the structure of the Chow ring of the (standard) compactification of $\mathcal{A}_3$ was determined in [25].

For $g = 2$ certain explicit compactifications for various moduli spaces of abelian surfaces have been studied in quite some detail, cf. [28].

An interesting problem form the $\mathbb{Q}$-rational points of $\mathcal{A}_g/\mathbb{Q}$. Since for $g \geq 7$ the moduli space is of general type one expects that the set of rational points is contained in a proper Zariski closed subset. It would be nice to have a candidate for this set; undoubtedly this is a difficult problem; compare the work of Mazur on the rational points of the modular curves $X_0(N)$ for $g = 1$. Note that $\mathcal{A}_g$ possesses lots of rational points: for example the closure of the locus of products of elliptic curves is a projective space $\mathbb{P}^g/\mathbb{Q}$, hence possesses many rational points. The Jacobians of curves of genus g defined over $\mathbb{Q}$ provide us also with many rational points of $\mathcal{A}_g$.

4   The Torelli Morphism

Let C be a (complete, absolutely irreducible, non-singular) algebraic curve over a field K. Its Jacobian variety $X = \mathrm{Pic}^0(C)$ has canonically a principal polarization (given by the "theta divisor"). In this way we obtain examples of principally polarized abelian varieties. In some sense these are the only examples that we know quite well because it is very difficult to describe abelian varieties by equations. Abelian varieties do not fit easily in projective space. For example, a principally polarized abelian variety of dimension g cannot be embedded in $\mathbb{P}^{2g}$ for $g \geq 2$, see [5]. They can be embedded in projective space using theta functions, but the dimension is quite high and the equations are complicated. Curves are much more easy to deal with and the geometric properties of Jacobians can be described in terms of the geometry of the curves. We think still the best way to describe abelian varieties is as quotients of Jacobians. In any case, it makes sense to compare the moduli of curves and of principally polarized abelian varieties. This is done via the map:

$$C \mapsto \mathrm{Jac}(C) = (\mathrm{Pic}^0(C), \lambda_C).$$

This defines a morphism between the coarse moduli schemes:

$$j : \mathcal{M}_g \to \mathcal{A}_g.$$

Comparing their dimensions we see that $\dim(\mathcal{M}_g) = \dim(\mathcal{A}_g) = g(g+1)/2$ if and only if $g \leq 3$. Hence we can expect that in these cases a principally polarized abelian variety is the generalized Jacobian of a curve; indeed this is true, as Weil proved for $g = 2$ and Oort and Ueno for $g = 3$, see [59]. However, for $g = 4$ the image has codimension 1. Still we can describe principally polarized abelian varieties of dimension 4 in terms of curves, namely as follows. Consider twofold étale covers $\pi : \tilde{C} \to C$ of curves of genus 5. Then the connected component of the kernel of the natural map $\pi_* : \operatorname{Jac}(\tilde{C}) \to \operatorname{Jac}(C)$ is a principally polarized abelian variety of dimension 4 and we can get all of them in this way. (Note that curves of genus 5 depend on 12 parameters, and principally polarized abelian varieties of dimension 4 depend on 10.) We come back to this in the next section.

In general the problem arises how to characterize the image of j. More precisely, the famous Torelli theorem says that this morphism is injective on geometric points:

Theorem 4.1. *(Torelli theorem.) Let k be an algebraically closed field. Then two algebraic curves C and C' over k are isomorphic if and only if their canonically polarized Jacobians are isomorphic:*

$$C \cong_k C' \Longleftrightarrow \operatorname{Jac}(C) \cong_k \operatorname{Jac}(C').$$

However this need not imply that the morphism j is an immersion. It turns out that this is indeed the case in characteristic zero, but in general it need not be; see [58].

Note that in the Torelli theorem we need not only the abelian variety, but also its polarization. For non-isomorphic curves with isomorphic abelian varieties, see [11], [27].

The Zariski closure of the image of j is called the *Torelli locus*:

$$\mathcal{T}_g := (j(\mathcal{M}_g))^{\operatorname{Zar}}.$$

The problem of characterizing Jacobians among all principally polarized abelian varieties, or in other words to characterize the closed subset $\mathcal{T}_g \subset \mathcal{A}_g$, is called the *Schottky problem* (although the problem was already signaled by Riemann). It is a difficult problem, which of course allows a priori several answers. Quite a lot of research has been devoted to it, see Mumford's survey [48], and also [6], [14], [23].

Schottky deduced relations that the image has to satisfy by using double covers of curves. It was proved by van Geemen (see [21]) that the Torelli locus is an irreducible component of the algebraic set described by these relations.

There are many other approaches to characterizing Jacobians among the principally polarized abelian varieties; they include for example an analysis of the singularities of the theta divisor (Andreotti-Mayer), reducibility of intersections $\Theta \cap \Theta_x$ (Weil) and the existence of trisecants (Gunning, Welters, Debarre).

Another approach due to van Geemen and van der Geer, and extended by Donagi, studies divisors in the linear system $|2\Theta|$, see [22], [15]. This linear system defines an embedding of the Kummer variety in projective space (if the abelian variety is not a product). By following the image of the origin for varying X we get a map from the moduli space to the same projective space. The idea is to compare the image of the moduli space and the Kummer variety. In this way one uses the geometry of the moduli space to formulate a (conjectural) answer to the Schottky problem

There is at least one conclusive good answer to the Schottky problem due to Shiota, [63]. He showed that complex abelian varieties which are Jacobians are characterized by the fact that their theta-function satisfies a 4-th order partial differential equation, the so-called Kadomtsev-Petviashvili-equation (or K-P). This relation comes from the fact that the Kummer variety of a Jacobian possesses a 4-dimensional family of trisecants (when embedded in projective space via $|2\Theta|$). A simpler proof, much more geometric in spirit, is due to Arbarello and De Concini, see [2]; see also [38]. This solved a conjecture of Novikov.

The various approaches are not unrelated. For example, if one studies infinitesimal versions of the approach using $|2\Theta|$ one arrives at Novikov's conjecture (or very close to it).

Finally, we would like to mention another aspect of abelian varieties that might lead to a characterization of Jacobians. Write a complex abelian variety X as V/Λ with $V \cong \mathbb{C}^g$ and Λ a lattice in V. The principal polarization defines a non-degenerate Hermitian form H on V. We now can look at the square of the minimal length of a non-zero lattice vector with respect to H:

$$m(X,\Theta) = \min_{x\in\Lambda, x\neq 0} H(x,x).$$

Buser and Sarnak (see [7]) have studied the maximum value M of $m(X,\Theta)$, where $m(X,\Theta)$ is viewed as a function on the moduli space. They showed that $M \geq \frac{1}{\pi}(2g!)^{1/g}$. Moreover, they show that for a Jacobian $X = \mathrm{Jac}(C)$ of a curve of genus g one has the upper bound

$$m(\mathrm{Jac}(C)) \leq \frac{3}{4}\log(4g+3).$$

This shows that Jacobians are special in the sense that they possess a very short period. Can we transform this into a good criterion for recognizing Jacobians ?

The invariant $m(X,\Theta)$ is closely related to another invariant, the Seshadri constant $\epsilon(X,\Theta,x)$ at any point x of X. This invariant measures the local positivity of the theta divisor at x. More precisely, by blowing up the variety X in x we get a new variety $\tilde{X}$ with a morphism $\pi : \tilde{X} \to X$ and an exceptional divisor E on $\tilde{X}$. Then one sets

$$\epsilon(X) = \sup\{\epsilon \geq 0 : \pi^*(\Theta) - \epsilon E \text{ is nef}\}.$$

This is independent of x and one has $\epsilon(X) \leq (g!)^{1/g}$. Lazarsfeld proved in [36] that $\epsilon(X) > (\pi/4)m(X)$.

In the simplest case, for $g = 4$, the Torelli locus has codimension 1 and can be described as the zero locus of an explicit modular form of weight 8, cf. [29], [22].

5   Cycles on the Moduli Space of Abelian Varieties

Knowledge of subvarieties in the moduli spaces of abelian varieties can tell us a lot about the geometry of these spaces. If one is able to calculate the cycle classes of such subvarieties, one obtains information about the cohomology of these spaces. There are several ways of constructing such subvarieties.

One way is to measure how far a principally polarized abelian variety is from being a Jacobian. If it is not a Jacobian then it could be the Prym variety of a double étale cover of a curve as we saw for $g = 4$. So we could look at the smallest genus such that X is a quotient of a Jacobian of that genus; in other words we look for the smallest genus of a curve on the abelian variety. This yields a "stratification" of the moduli space $\mathcal{A}_g$, where the smallest stratum is the Torelli locus. But we do not know much about this stratification. For example, let $\gamma = \gamma(g)$ be the minimum genus of a curve on a generic principally polarized abelian variety of dimension g. Then one can prove that for $g \geq 4$ one has $\gamma \geq g(g - 1)/2 + 1$, see [57]. It is known that $\gamma(4) = 7$ (see [4]) and $\gamma(5) = 11$ and these are the only two cases where the bound is attained. There is a characterization of the abelian varieties of dimension $g = 4$ (resp. $g = 5$) having a curve of genus $\gamma(4)$ (resp. $\gamma(5)$); for the $g = 4$ case, see [4], and for $g = 5$ see [57]. As to the cycle classes even less seems known. For example, one does not even know the cycle class of the Torelli locus. Faber has calculated the tautological part of it, cf. [18].

A related but slightly different approach studies the existence of curves of given degree m on an abelian variety: $C \cdot \Theta = m$. See [13]. Or one tries to determine the loci of abelian varieties for which the integral class $d\Theta^{g-1}/(g-1)!$ is represented by an effective curve. For $d = 1$ one finds Jacobians by the Matsusaka-Ran criterion; Prym varieties yield $d = 2$ and are characterized by this, cf. [66].

Another thing one can do is to look at the endomorphism rings of principally polarized abelian varieties. For example, if we work over $\mathbb{C}$, then the generic abelian variety has $\mathrm{End}(X) = \mathbb{Z}$ and for a simple abelian variety (i.e. one that is not isogenous to a product) the richest endomorphism ring is an order in a so-called CM-field (a totally imaginary quadratic extension K of a totally real field) with $\dim(X) = 2 \deg[K : \mathbb{Q}]$.

Over $\mathbb{C}$ the moduli spaces of principally polarized abelian varieties with endomorphism ring of a certain type (or larger) are described as quotients of bounded symmetric domains. The corresponding subvarieties of $\mathcal{A}_g$ are described by a map between bounded symmetric domains

$$\mathfrak{g} \to \mathfrak{h}_g;$$

the image of such a map is a totally geodesic subspace in $\mathfrak{h}_g$. In some sense we can describe them quite well, cf. Satake's book [61]. On the other hand it becomes very difficult to compare them with the Torelli locus. We give an example.

Definition. A simple abelian variety X is called of CM-type (better: admits sufficiently many complex multiplications) if its endomorphism algebra $\text{End}^\circ(X) = \text{End} \otimes_{\mathbb{Z}} \mathbb{Q}$ contains a field of degree $2g$ over $\mathbb{Q}$. For a simple abelian variety over $\mathbb{C}$ this is the same as requiring that $L = \text{End}^\circ(X)$ is a field of degree $2g$ over $\mathbb{Q}$; in that case L is a purely imaginary extension of a totally real field of degree g over $\mathbb{Q}$. An abelian variety X admits sufficiently many complex multiplications if it is isogenous with a product of simple abelian varieties having this property; equivalently: if $\text{End}^\circ(X)$ contains a commutative semi-simple algebra of rang $2 \cdot \dim(X)$. A point in $\mathcal{A}_g(\mathbb{C})$ is called a CM-point if the corresponding abelian variety is of CM-type.

Coleman posed the problem whether there are many CM curves, i.e. curves whose Jacobian is of CM-type. It is easy to see that for $g \leq 3$ there are infinitely many CM curves of genus g. De Jong and Noot proved in [30] the same for $g = 4$ and $g = 6$. It is an open problem what the answer to the question posed by Coleman is for other values of g, i.e. whether the number of CM curves of a given genus over $\mathbb{C}$ is finite or infinite.

On the other hand, there is a result of Ciliberto, van der Geer and Teixidor i Bigas ([9], [10]) which says that for $g \geq 1$ a subvariety of $\mathcal{M}_g$ whose generic point parametrizes simple Jacobians with $\text{End}(J) \neq \mathbb{Z}$ has codimension $\geq g - 1$. Note that for $\mathcal{A}_g(\mathbb{C})$ there is the same bound for the codimension of subvarieties classifying abelian varieties with non-trivial endomorphism ring (i.e. $\neq \mathbb{Z}$). However, one expects excess intersection of the Torelli locus and the loci corresponding to abelian varieties with *very large* endomorphism rings; that is, one expects that they intersect much more than their dimensions suggest.

Yet another approach is more geometric. One looks at the theta divisor of the abelian variety and at its singular locus. The generic principally polarized abelian variety has a smooth theta divisor. Those abelian varieties whose theta divisor is singular form a divisor in $\mathcal{A}_g$ as was shown by Beauville and Mumford, cf. [47]. More precisely, we can define following Andreotti-Mayer the following loci:

$$N_k = \{[(X, \Theta)] \in \mathcal{A}_g : \dim \text{Sing}(\Theta) \geq k\}.$$

Mumford [47] calculated the cycle class of N_0, and he proved that $\text{codim}(N_k) \geq 2$ for $k \geq 1$; but it seems difficult to say much about the codimension and the components of the N_k for $k \geq 1$. Recently, Ciliberto and van der Geer proved that for $g \geq 2$, $k \geq 1$ one has $\text{codim}(N) \geq k + 2$ for an irreducible component N of $N_{\tilde{k}}$ whose generic point classifies a simple abelian variety, [12].

In contrast to several other approaches here we can say something if we compare the Torelli locus with the cycles N_k and that is why these cycles were introduced by Andreotti and Mayer. They proved that $\mathcal{T}_g$ is an irreducible component of N_{g-4} for $g \geq 4$ and this constitutes one of the approaches to the Schottky problem, cf. [48].

6   The Moduli Space $\mathcal{A}_g$ in Positive Characteristic

The moduli spaces of abelian varieties in characteristic $p > 0$ possess properties which do not occur in characteristic 0. These phenomena are mainly due to the occurrence of inseparable maps and of finite group schemes that are not reduced. To the mathematician used to work in characteristic 0 these properties might be unfamiliar and strange. But the specific characteristic p properties give us insights that are not available in characteristic 0. As an example one might take the existence of complete subvarieties as mentioned above.

This may motivate us to study the specific structure of the moduli spaces in characteristic p and exploit the characteristic p phenomena there. It turns out that this can help us to understand the moduli spaces in characteristic 0 as well. For some people that might be a reason to study characteristic p.

The idea to use characteristic 0 to deduce results in characteristic p is of course an old one. One studies geometric questions by first deducing properties over the complex numbers and then one tries to extend these over a base ring like $\mathbb{Z}$ and to reduce modulo p. For example, in this way (by a procedure employed by Zariski, Grothendieck and many others) the irreducibility of $\mathcal{M}_g \otimes K$ for any field K was proved by Deligne and Mumford (starting from the classical fact that it is known in case $K = \mathbb{C}$). Analogously, the irreducibility of $\mathcal{A}_g \otimes K$ for any field K was proved by Faltings and by Chai.

To give an example of a specific characteristic p phenomenon, consider the multiplication by p map for an abelian variety X. Its degree of inseparability is p^d with $d = d(X)$ an integer between g and $2g$. Equivalently, consider the kernel of multiplication by p as set of points, and define the p-rank, $f(X)$ of X by:

$$f(X) = f \iff \#X[p](k) = p^f,$$

where k is an algebraically closed field containing K. We have $f(X) = 2g - d(X)$. It turns out that all values $0 \leq f \leq g$ appear.

We say that an abelian variety X with $f(X) = g = \dim(X)$, is *ordinary*; experience shows that ordinary abelian varieties in positive characteristic behave very much like abelian varieties in characteristic zero. However non-ordinary abelian varieties have a competely different flavour and are more similar to degenerate (semi-stable) abelian varieties in characteristic 0.

By looking at the p-rank we obtain a stratification of $\mathcal{A}_g \otimes \mathbb{F}_p$ by closed subsets V_f where the p-rank is at most f. Their codimension is $g - f$ as shown by Koblitz in the principally polarized case and proved by Norman and Oort in the general case, see [34], [51]. We see that already this invariant $f = f(X)$ provides us with valuable information. Note that this information "changes from fibre to fibre" in $\mathcal{A}_g \to \operatorname{Spec}(\mathbb{Z})$. This is a phenomenon which should be studied further, and which promises new results and views on the structure of moduli spaces.

One can view a change of this structure as a kind of "degeneration", where eventually very special abelian varieties (called *superspecial*; they are products of supersingular elliptic curves) are the zero dimensional strata.

In fact, if an abelian variety X degenerates (specializes to a semi-abelian variety X_0 which is not an abelian variety), then it loses torsion points: as a group $X[n](k) \cong (\mathbb{Z}/n)^{2g}$ and $X_0[n](k) \cong (\mathbb{Z}/n)^m$ with $m < 2g$; here k is an algebraically closed field, and n is an integer prime to the characteristic of the base field.

If X is defined in characteristic $p > 0$, and X is ordinary and X_0 is not ordinary, then $X[p](k) \cong (\mathbb{Z}/p)^g$ and $X_0[p](k) \cong (\mathbb{Z}/p)^f$ with $f < g$.

In stratifications of $\mathcal{A} \otimes \mathbb{F}_p$ this aspect is worked out in detail. Not only the p-rank is considered, but finer invariants are studied, and we obtain the following stratifications.

The Ekedahl-Oort-stratification. This stratification, an extension of the stratification by p-rank, was constructed by Ekedahl and Oort, cf. [17], [54]. For an abelian variety X one considers the kernel $X[p]$ of multiplication by p. This is a group scheme of order p^{2g}. It turns out that this group scheme, with the pairing induced from a principal polarization on X, is already defined over $\mathbb{F}_p$. In particular there are only finitely many possibilities, each giving rise to a locally closed subset of $\mathcal{A} = \mathcal{A}_g \otimes \mathbb{F}_p$. In this way we obtain the EO-*stratification*. Using an idea by Raynaud, plus positivity of the determinant of the Hodge bundle (as proved by Moret-Bailly, see [44]) one proves that these strata are quasi-affine. We obtain a cell-like decomposition of $\mathcal{A}$. Interestingly, the combinatorics of this does not depend on p, but the Chow classes of the strata do depend on p. Note that $X[p]$ in general is not invariant under isogenies. The number of strata in the EO-stratification is equal to 2^g, the strata cover all dimensions between 0 and $g(g + 1)/2$. The dimension of each stratum is easily computed from the combinatorics used in the classification of $X[p]$ but this fact is not so easily proved.

This stratification is explained by Oort in one of the papers in this volume ([54]). In another paper [24] the cycle classes of this stratification are computed (joint work of Ekedahl and van der Geer). It turns out that these classes all lie in the subring of the Chow ring (or cohomology ring) generated by the Chern classes of the Hodge bundle. As an example, consider the locus V_f of abelian varieties whose p-rank is $\le f$.

Theorem 6.1. *The class of the locus V_f of abelian varieties with p-rank $\leq f$ can be written as*

$$[V_f] = (p-1)(p^2-1)\ldots(p^{g-f}-1)\,\lambda_{g-f},$$

with λ_{g-f} the $(g-f)$-th Chern class of $\mathbb{E}$.

This formula generalizes the classical formula of Deuring for the number of supersingular elliptic curves given above, see [24]. This formula has interesting consequences; for example, the positivity properties of the Chern classes λ_i in characteristic p follow from it. Some of the strata can be described very explicitly.

It seems again difficult to compare this stratification with the Torelli locus. For some partial results we refer to [19].

The dichotomy 'ordinary vs. supersingular' for $g = 1$ can be generalized in at least two rather different ways. One of these leads to the EO-stratification. This stratification is not stable with respect to isogenies. The other generalization is the stratification obtained by looking at the p-divisible group of an abelian variety as follows.

The Newton polygon-stratification. For this stratification we do not look at the kernel $X[p]$ of multiplication by p alone, but consider the asymptotic properties of $X[p^n]$ as $n \to \infty$. A p-divisible group G over a field $K \supset \mathbb{F}_p$ defines a Newton polygon, denoted by $\mathcal{N}(G)$. If G is defined over a finite field, the Newton polygon can be defined using the geometric Frobenius endomorphism. Namely, consider the characteristic polynomial of the Frobenius endomorphism (acting on the Tate-module T_ℓ for $\ell \neq p$) and its Newton polygon. For the general case one uses Dieudonné module theory for the definition. Note that the notion of Newton polygon also refines the notion of p-rank; this new notion is isogeny invariant. This defines a stratification on $\mathcal{A}$, the largest stratum being the set of all ordinary principally polarized abelian varieties, the smallest stratum being the supersingular ones. The fact that these strata are locally closed was proved by Grothendieck and Katz. The Newton polygon 'goes up' under specialization. This stratification is difficult to describe in local deformation theory (because Newton polygons depend on the whole p-divisible group). It has been proved, see [31], that if in a stratum the Newton polygon changes the locus where it does so has codimension 1. This stratification is the natural one when Hecke orbits are concerned because of the invariance of Newton polygons under isogenies, cf. Chai's work on Hecke orbits [8]. The smallest stratum, the supersingular one, has dimension equal to $[g^2/4]$, as was proved by Li and Oort [37]. The dimension of each stratum can be easily expressed in the combinatorics of the Newton Polygon diagram; for the statement and the proofs, see [53], [55], [56].

The cycle classes are not known in general. In a few cases they can be computed, e.g. for the supersingular locus for $g = 3$ and the tautological part also for $g = 4$, cf. [24].

Again, not too much is known about the intersection of this stratification with the Torelli locus. For example, it is not known whether there exists for every g in characteristic $p > 2$ a curve of genus g whose Jacobian is supersingular. In characteristic 2 however, this is known and there is an explicit construction of such curves, see [26].

7 The Moduli Space in Mixed Characteristic

The study of algebraic geometry in mixed characteristics is a fascinating topic in mathematics. It is a field where analysis and algebra meet, number theory and geometry each find answers for the other's difficult problems. As a good example of a beautiful result we mention the theorem of Elkies [16] saying that for an elliptic curve $E/\mathbb{Q}$ there are infinitely many primes p such that $E \bmod p$ is supersingular. It is an interesting challenge to extend this result to abelian varieties of higher dimensions.

We have seen that the moduli spaces $\mathcal{M}_g$ for algebraic curves, and $\mathcal{A}_g$ for abelian varieties (and their compactifications) are defined over $\mathrm{Spec}(\mathbb{Z})$. The scene is set to apply p-adic methods, to compare topological-analytical methods with modular techniques, and to turn back and forth from one method to another in order to answer the questions we consider. We mention one case in which comparisons give concrete answers and lead to many questions.

7.1 Serre-Tate coordinates: canonical coordinates on $\mathcal{A}_g$

Let us consider $\mathcal{A}_g$, and consider a field K of characteristic p and a point $x \in \mathcal{A}_g(K)$, corresponding to an ordinary polarized abelian variety. Serre and Tate have constructed "canonical coordinates" on the completion of $\mathcal{A}_g$ at the point x. These coordinates in mixed characteristics turn out to have the property that subvarieties of Hodge type are characterised as those algebraic varieties in $\mathcal{A}_g$ which in at least one (and hence in all) ordinary point are linear in the Serre-Tate coordinates (as was proved by Noot, and by Moonen; see [42]). Moreover, such Shimura varieties can be considered as subsets of $\mathcal{A}_g(\mathbb{C})$ and a subvariety is of Hodge type if and only if it contains a CM-point, and it is totally geodesic in the metric coming from the Siegel upper half space (see [41]). For further details and references, see [43]. We see a fascinating relation between structures coming from complex-analytic considerations, from p-adic techniques, and from characteristic p considerations. We expect many more implications of these interesting new ideas.

7.2   Indigenous bundles: canonical coordinates on $\mathcal{M}_g$

At the same time, and completely independently, Mochizuki considered algebraic curves in mixed characteristics. An idea of Gunning from 1967 inspired Mochizuki to define canonical coordinates on $\mathcal{M}_g$. The essential idea is that a hyperbolic algebraic curve (for example a complete, non-singular irreducible curve of genus $g \geq 2$) is parametrized by the upper half plane; this gives a PSL(2,$\mathbb{R}$) representation of its fundamental group, which by a GAGA-type of principle, gives rise to an algebraic $\mathbb{P}^1$-bundle. This seemingly analytic parametrization leads to an algebraic description, and Mochizuki uses this idea to construct "canonical coordinates" locally on $\mathcal{M}_g$ in the p-adic sense (see [40]). Moreover these coordinates have their counterpart coming from the Kähler metric on the moduli spaces considered.

The analogy between the two considerations is striking. However, it seems extremely difficult to compare these two aspects under the Torelli morphism. When are the canonical coordinates locally on $\mathcal{M}_g$ under the morphism j part of the Serre-Tate coordinates on $\mathcal{A}_g$?

A further study of these aspects seems challenging, difficult, and it might lead to fascinating and new mathematics.

Bibliography

[1] V. Alexeev: *Complete moduli in the presence of semi-abelian group action.* Preprint July 1998.

[2] E. Arbarello, C. De Concini: *Another proof of a conjecture of S.P. Novikov on periods of Abelian integrals on Riemann surfaces.* Duke Math. J. **54** (1987), no. 1, 163–178.

[3] W. Baily, A. Borel: *Compactification of arithmetic quotients of bounded symmetric domains.* Annals of Math. **84** (1966), 442–528.

[4] F. Bardelli, C. Ciliberto, A. Verra: *Curves of minimal genus on a general abelian variety.* Compositio Math. **96** (1995), no. 2, 115–147.

[5] W. Barth, A. Van de Ven: *On the embedding of Abelian varieties into projective spaces.* Ann. Mat. pura appl. **103** (1975), 127–129.

[6] A. Beauville: *Le problème de Schottky et la conjecture de Novikov.* Séminaire Bourbaki, Vol. 1986/87. Astérisque No. 152-153 (1987), 4, 101–112 (1988).

[7] P. Buser, P. Sarnak: *On the period matrix of a Riemann surface of large genus.* Invent. Math. **117** (1994), 27–56.

[8] C.L. Chai: *Every ordinary symplectic isogeny class in positive characteristic is dense in the moduli.* Invent. Math. **121** (1995), 439–479.

[9] C. Ciliberto, G. van der Geer, M. Teixidor i Bigas: *On the number of parameters of Jacobians which possess non-trivial endomorphisms.* J. Algebraic Geom. **1** (1992), 215–229.

[10] C. Ciliberto, G. van der Geer: *Subvarieties of the moduli space of curves parametrizing Jacobians with non-trivial endomorphisms.* Amer. J. of Math. **114** (1991), 551–570.

[11] C. Ciliberto, G. van der Geer: *Non-isomorphic curves of genus four with isomorphic (non-polarized) jacobians.* In: *Classification of Algebraic Varieties* (Eds. C. Ciliberto, E.L. Livorni, A.J. Sommese), Contemporary Math. **162** (1994), 129–133.

[12] C. Ciliberto, G. van der Geer: Manuscript in preparation.

[13] O. Debarre: *Degrees of curves in abelian varieties.* Bull. Soc. Math. France **122** (1994), no. 3, 343–361.

[14] O. Debarre: *The Schottky problem: an update.* Current topics in complex algebraic geometry (Berkeley, CA, 1992/93) (Eds. C.H. Clemens and J. Kollár), 57–64, Math. Sci. Res. Inst. Publ., 28, Cambridge Univ. Press, Cambridge, 1995.

[15] R. Donagi: *The Schottky Problem.* Theory of moduli (Montecatini Terme, 1985) (Ed. E. Sernesi), 84–137, Lecture Notes in Math. **1337**. Springer, Berlin-New York, 1988.

[16] N. Elkies: *The existence of infinitely many supersingular primes for every elliptic curve over* $\mathbb{Q}$, Invent. Math. **89** (1987), 561–567.

[17] T. Ekedahl, F. Oort: *A stratification of a moduli space of abelian varieties.* In preparation.

[18] C. Faber: *Algorithms for computing intersection numbers on moduli spaces of curves, with an application to the class of the locus of Jacobians*, in *Algebraic Geometry* (Eds. F. Catanese, K. Hulek, C. Peters, M. Reid), Cambridge University Press, 1999.

[19] C. Faber, G. van der Geer: *Complete subvarieties of moduli spaces and the Prym map.* In preparation.

[20] G. Faltings, C.-L.Chai: *Degeneration of abelian varieties.* Ergebnisse 3.22, Springer, Berlin-New York, 1990.

[21] B. van Geemen: *Siegel modular forms vanishing on the moduli space of curves.* Inv. Math. **78** (1984), 329–349.

[22] B. van Geemen, G. van der Geer: *Kummer varieties and the moduli spaces of abelian varieties.* Amer. J. Math. 108 (1986), no. 3, 615–641.

[23] G. van der Geer: *The Schottky Problem.* In: Arbeitstagung Bonn 1984, 385–406, Lecture Notes in Math. **1111**, Springer, Berlin-New York, 1985.

[24] G. van der Geer: *Cycles on the moduli space of abelian varieties.* This volume, 65–89.

[25] G. van der Geer: *The Chow ring of the moduli space of abelian threefolds*, J. Algebraic Geom. **7** (1998), 753–770.

[26] G. van der Geer, M. van der Vlugt: *On the existence of supersingular curves of given genus*, J. Reine Angew. Math. **45** (1995), 53–61.

[27] E. Howe: *Constructing distinct curves with isomorphic Jacobians.* J. Number Theory **56** (1996), no. 2, 381–390.

[28] K. Hulek, C. Kahn, S.H. Weintraub: Moduli spaces of abelian surfaces: compactification, degenerations, and theta functions. De Gruyter Expositions in Mathematics, 12. Walter de Gruyter & Co., Berlin, 1993.

[29] J.I. Igusa: *On the irreducibility of Schottky's divisor.* J. Fac. Sci. Tokyo **28** (1981), 531–545.

[30] J. de Jong, R. Noot: *Jacobians with Complex Multiplication.* In: Arithmetic Algebraic Geometry (Eds. G. van der Geer, F. Oort, J. Steenbrink), 177–192, Progress in Math. **129**, Birkhäuser, Boston, 1991.

[31] J. de Jong, F. Oort: *Purity of the stratification by Newton polygons.* Utrecht Preprint 1081, December 1998; 35 pp.

[32] P. Katsylo: *Rationality of the moduli variety of curves of genus 3.* Comment. Math. Helv. **71** (1996), 507–524.

[33] N. Katz, B. Mazur: *Arithmetic moduli of elliptic curves.* Annals of Math. Studies **108**, Princeton University Press 1985.

[34] N. Koblitz: *p-adic variation of the zeta function over families of varieties over finite fields.* Compositio Math. **31** (1975), 119–218.

[35] H. Lange, Ch. Birkenhake: *Complex abelian varieties.* Grundlehren der Mathematischen Wissenschaften, 302. Springer, Berlin-New York, 1992.

[36] R. Lazarsfeld: *Lengths of periods and Seshadri constants of abelian varieties.* Math. Res. Lett. **3** (1996), no. 4, 439–447.

[37] K.Z. Li, F. Oort: *Moduli of supersingular abelian varieties.* Lecture Notes in Mathematics **1680**, Springer, Berlin-New York, 1998.

[38] G. Marini: *A geometrical proof of Shiota's theorem on a conjecture of Novikov.* Compositio Math. **111** (1998), 305–322.

[39] B. Mazur: *Rational points on $X_0(N)$.* Modular Functions of One Variable V, 107–148, Lecture Notes in Math. **601**, Springer, Berlin-New York, 1977.

[40] S. Mochizuki: *A theory of ordinary p-adic curves.* Publ. Res. Inst. Math. Sci. 32 (1996), no. 6, 957–1152.

[41] B. Moonen: *Linearity properties of Shimura varieties, I.* J. Algebraic Geom. **7** (1998), 539–567.

[42] B. Moonen: *Linearity properties of Shimura varieties, II.* Compositio Math. **114** (1998), 3–35.

[43] B. Moonen: *Models of Shimura varieties in mixed characteristics.* In: Galois representations in arithmetic algebraic geometry (Ed. A. J. Scholl & R. L. Taylor). London Math. Soc. Lect. Notes Series **254**, Cambridge Univ. Press, Cambridge 1998; p. 267 - 350.

[44] L. Moret-Bailly: *Pinceaux de variétés abéliennes.* Astérisque 129, Soc. Math. France, 1985.

[45] D. Mumford, J. Fogarty, F. Kirwan: *Geometric invariant theory.* Ergebnisse 34, Springer, Berlin-New York, 1994.

[46] D. Mumford: *Abelian Varieties.* Oxford University Press, 1974.

[47] D. Mumford: On the Kodaira dimension of the Siegel modular variety. Algebraic geometry—open problems (Ravello, 1982) (Eds. C. Ciliberto, F. Ghione, F. Orecchia), 348–375, Lecture Notes in Math. **997**, Springer, Berlin-New York, 1983.

[48] D. Mumford: *Curves and their Jacobians*. University of Michigan Press, Ann Arbor 1975.

[49] D. Mumford: *Hirzebruch's proportionality theorem in the noncompact case*. Invent. Math. **42** (1977), 239–272.

[50] D. Mumford: *Towards an enumerative geometry of the moduli space of curves*. In: Arithmetic and Geometry (Eds. M. Artin and J. Tate), Vol. II, 271–328, Progress in Math. **36**, Birkhäuser, Boston, 1983.

[51] P. Norman, F. Oort: *Moduli of abelian varieties*. Annals of Math. **112** (1980), 413–439.

[52] F. Oort: *Subvarieties of moduli spaces*. Inv. Math. **24** (1974), 95–119.

[53] F. Oort: *Moduli of abelian varieties and Newton Polygons*. Compt. Rend. Acad. Sc. Paris **312** Sér. I (1991), 385–389.

[54] F. Oort: *A stratification of a moduli space of polarized abelian varieties in positive characteristic*. This volume, 47–64.

[55] F. Oort: *Newton polygons and formal groups: conjectures by Manin and Grothendieck*. Utrecht Preprint 995, January 1997.

[56] F. Oort: *Newton polygon strata in the moduli space of abelian varieties*. In preparation.

[57] G.P. Pirola: *Abel-Jacobi invariant and curves on generic abelian varieties*. In: Abelian varieties (Egloffstein, 1993) (Eds. W. Barth, K. Hulek and H. Lange), 237–249, de Gruyter, Berlin, 1995.

[58] F. Oort, J. Steenbrink: *The local Torelli problem for algebraic curves*. In: Algebraic Geometry (Angers, 1979) (Ed. A. Beauville), 157–204, Sijthoff & Noordhoff, 1980.

[59] F. Oort, K. Ueno: *Principally polarized abelian varieties of dimension two or three are Jacobian varieties*. J. Fac. Sci. Univ. Tokyo Sect. IA Math. **20** (1973), 377–381.

[60] I. Satake: *On the compactification of the Siegel space*. J. Indian Math. Soc. **20**, 259–281 (1965).

[61] I. Satake: *Arithmetic structures of symmetric domains*. Iwanami Shoten/ Princeton University Press, 1980.

[62] G. Shimura: *Introduction to the arithmetic theory of automorphic forms*. Princeton University Press, 1971.

[63] T. Shiota: *Characterization of Jacobian varieties in terms of soliton equations*. Inv. Math. **83** (1986), 333–382.

[64] J. Silverman: *Arithmetic of elliptic curves*. Springer, Berlin-New York, 1986.

[65] G. Welters: *A criterion for Jacobi varieties*. Ann. of Math. **120** (1984), 497–504.

[66] G. Welters: *Curves of twice the minimal class on principally polarized abelian varieties*. Nederl. Akad. Wetensch. Indag. Math. **49** (1987), 87–109.

Remarks on Moduli of Curves

Carel Faber and Eduard Looijenga

Abstract

We discuss some aspects of the theory of the moduli space of curves as well as some recent research.

1 Introduction

Rather than trying to provide here a comprehensive introduction to moduli of curves, we have chosen to limit the discussion to certain aspects of the theory. We also survey some of the recent research directly related to the papers in this volume.

After introducing the mapping class group and the Torelli group, the moduli space of curves is constructed as an analytic orbifold. We discuss the Deligne-Mumford-Knudsen compactification and the fact that there exist global smooth covers of it. Next we recall the definition of the tautological classes and the results on the stability of the homology of the mapping class group as well as Mumford's conjecture. Sections 10 and 13 discuss the Witten conjecture that was proved by Kontsevich and its generalization to moduli spaces of stable maps. We conclude with a discussion of complete subvarieties of moduli space and of recent results regarding its intersection theory.

For more information on the moduli space of curves, the reader could consult, e.g., the books [62, 63, 52, 45], the collections [10, 17], and the survey papers [44, 40].

2 Mapping Class Groups

Fix a closed connected oriented surface S_g of genus g and a sequence of distinct points $x_0, x_1, \ldots$ on S_g and let us write $S_{g,n}$ for $S_g - \{x_1, \ldots, x_n\}$ and $\pi_{g,n}$ for $\pi_1(S_{g,n}, x_0)$. If the subscript n is omitted, it is assumed to be zero. We stick to this notation throughout this introduction.

We begin with noting that in the absence of any punctures we have a natural isomorphism $H_2(\pi_g; \mathbb{Z}) \cong H_2(S_g, \mathbb{Z})$, so that the orientation of S_g defines a distinguished generator of $H_2(\pi_g; \mathbb{Z})$. For positive n, the simple positively oriented loops around x_i make up a distinguished conjugacy class B_i in $\pi_{g,n}$, $i = 1, \ldots, n$. There is a standard way to present the group $\pi_{g,n}$ with generators $\alpha_{\pm 1}, \ldots, \alpha_{\pm g}, \beta_1, \ldots, \beta_n$, subject to the relation

$$(\alpha_1, \alpha_{-1}) \cdots (\alpha_g, \alpha_{-g}) \beta_n \cdots \beta_1 = 1,$$

where in case $n = 0$, the generators $\alpha_{\pm 1}, \ldots, \alpha_{\pm g}$ have been chosen compatibly with the orientation, and $\beta_i \in B_i$ in case $n > 0$. In the latter case, $\pi_{g,n}$ is just a free group on $2g + n - 1$ generators and it is the data of $B_1, \ldots, B_n$ that give this group its extra structure. The inclusion $S_{g,n+1} \subset S_{g,n}$ induces a surjective homomorphism $\pi_{g,n+1} \to \pi_{g,n}$ on fundamental groups and we can arrange that the generators of $\pi_{g,n}$ are the images of their namesakes in $\pi_{g,n+1}$.

One defines the *n-pointed mapping class group of genus g*, here denoted by Γ_g^n, as the connected component group of the group of orientation preserving self-homeomorphisms of S_g that fix $x_1, \ldots, x_n$. This group acts by outer automorphisms (that is, the action is given up to inner automorphisms) on the fundamental group $\pi_{g,n}$. It is a classical result that this action is faithful. According to Nielsen and Zieschang [76] the image can be characterized as follows: for $n = 0$, $\Gamma_g = \Gamma_g^0$ maps onto the group of outer automorphisms of π_g which act trivially on $H_2(\pi_g; \mathbb{Z})$, and for $n > 0$, Γ_g^n maps onto the group of outer automorphisms of $\pi_{g,n}$ which preserve each conjugacy class B_i, $i = 1, \ldots, n$. So this yields a description of the mapping class group purely in terms of group theory. Forgetting x_{n+1} defines an obvious homomorphism $\Gamma_g^{n+1} \to \Gamma_g^n$. Since any orientation preserving self-homeomorphism of S_g that fixes $x_1, \ldots, x_n$ is isotopic to one that also fixes x_{n+1}, this homomorphism is surjective. An arc in $S_{g,n}$ connecting x_0 and x_{n+1} allows us to identify Γ_g^{n+1} with the group of automorphisms of $\pi_1(S_{g,n}, x_{n+1}) \cong \pi_{g,n}$ which preserve the conjugacy classes $B_1, \ldots, B_n$. With this identification, the kernel of $\Gamma_g^{n+1} \to \Gamma_g^n$ is the group of inner automorphisms of $\pi_{g,n}$. If we take $g = n = 1$, then we get the familiar identification of Γ_1^1 as the group of orientation preserving automorphisms of $\pi_{1,0} \cong \mathbb{Z}^2$: $\Gamma_1^1 \cong \mathrm{SL}(2, \mathbb{Z})$.

A special set of elements of the mapping class group Γ_g^n are the Dehn twists. A *Dehn twist* is given by a regularly embedded circle $\delta \subset S_{g,n}$ and then represented by a homeomorphism which on a closed neighborhood of that circle (with oriented parameterization by $(\phi, t) \in S^1 \times [-1,1]$) equals $(\phi, t) \mapsto (\phi + \pi(t+1), t)$ and is the identity outside that neighborhood. The corresponding element of Γ_g^n, denoted by τ_δ, only depends on the isotopy class of δ. The Dehn twist τ_δ is the identity element of Γ_g^n if and only if δ bounds a disk on S_g which contains at most one x_i, $i = 1, \ldots, n$; for that reason such a circle is called *trivial* also. It is known that finitely many Dehn twists generate the whole mapping class group. For the unpunctured case $n = 0$ a relatively simple finite presentation of Γ_g (with Dehn twist generators) has been given by Wajnryb [72], and for the general case one was recently obtained by Gervais [29].

Although mapping class groups have been studied since their introduction some seventy years ago, they are still mysterious in many ways. In certain regards they behave as if they were arithmetic groups, but as Ivanov has shown, they are, apart from a few exceptions, not isomorphic to such a group. Following Hain and Morita, a mapping class group can however be naturally embedded in a proarithmetic group (its proarithmetic hull), and the latter is at present much better accessible. We discuss this briefly in section 9.

3   The Torelli Group

We here focus on the unpunctured case: $n = 0$. The homology group $H_1(S_g; \mathbb{Z})$ is then of rank $2g$ and the orientation equips it with a unimodular symplectic form. Let us write V_g for $H_1(S_g; \mathbb{Z})$ and $\omega_g \in \wedge^2 V_g$ for the 'inverse' symplectic form. So if $a_i \in V_g$ is the class of the generator α_i of π_g, then $a_{\pm 1}, \ldots, a_{\pm g}$ is a basis for V_g and $\omega_g = a_1 \wedge a_{-1} + \cdots + a_g \wedge a_{-g}$. It is clear that the mapping class group Γ_g acts on V_g and leaves ω_g invariant. The image of this representation is in fact the full integral symplectic group $\mathrm{Sp}(V_g)$. Its kernel is called the *Torelli group of genus g*, denoted T_g. This group is trivial for genus zero and one; so that for instance $\Gamma_1 \cong \mathrm{Sp}(V_1) \cong \mathrm{SL}(2, \mathbb{Z})$. But for $g \geq 2$ the Torelli group contains the Dehn twists around circles which separate S_g into two connected components and for $g \geq 3$, also the elements of the form $\tau_\delta \tau_{\delta'}^{-1}$, where δ and δ' are disjoint circles on S_g, which together separate S_g into two pieces. According to Powell these elements generate T_g. The Dehn twists around separating circles generate a subgroup K_g of T_g that is clearly normal in Γ_g. We have $K_2 = T_2$. In fact, G. Mess showed that K_2 is the free group on the separating Dehn twists and that these generators are in bijective correspondence with the symplectic splittings of V_2 into two copies of V_1, hence infinite in number. The situation is quite different when $g \geq 3$. Dennis Johnson, who in the early eighties began a systematic study of the Torelli group, showed that T_g is then finitely generated and exhibited a remarkable epimorphism of T_g onto the lattice $\wedge_o^3 V_g := \wedge^3 V_g / (V_g \wedge \omega_g)$ with kernel K_g. An explicit (but perhaps not very insightful) way to describe this epimorphism is to say what it does to an element $\tau_\delta \tau_{\delta'}^{-1}$ as above: let $S \subset S_g$ be a connected component of $S_g - \delta - \delta'$ and orient δ as boundary component of S, so that it determines a class $d_S \in V_g$. The image V_S of $H_1(S; \mathbb{Z})$ in V_g is a sublattice on which the symplectic form is degenerate with kernel spanned by d_S. The form on $V_S / \mathbb{Z} d_S$

is unimodular and thus defines an element $\omega_S \in \wedge^2(V_S/\mathbb{Z}d_S)$. We can regard $d_S \wedge \omega_S$ as an element of $\wedge^3 V_g$. If S' is the other component of $S_g - \delta - \delta'$, then $d_{S'} = -d_S$ and $d_S \wedge (\omega_S + \omega_{S'}) = d_S \wedge \omega_g$. So the image of $d_S \wedge \omega_S$ in $\wedge_o^3 V_g$ only depends on the ordered pair (δ, δ'). This is the image of $\tau_\delta \tau_{\delta'}^{-1}$ under Johnson's homomorphism. A more natural description will be given in section 9.

4   Moduli Spaces of Curves

Let us now assume that g and n are such that the Euler characteristic $2-2g-n$ of $S_{g,n}$ is negative, in other words, exclude the cases of genus zero with at most two punctures and genus one without punctures. Then the set of complex structures on S_g compatible with the given orientation and given up to isotopy relative $x_1, \ldots, x_n$ is in a natural way a complex manifold of complex dimension $3g-3+n$. This manifold, which we shall denote $\mathcal{T}_{g,n}$, is called the *n-pointed Teichmüller space of genus g*. It is known that $\mathcal{T}_{g,n}$ is contractible and isomorphic to a bounded domain. Notice that there is an evident action of the mapping class group Γ_g^n on $\mathcal{T}_{g,n}$. This action is faithful and properly discrete, and so the orbit space has the structure of an *analytic orbifold*. As such it is denoted by $\mathcal{M}_{g,n}$. From the definition it is clear that the points of $\mathcal{M}_{g,n}$ are in bijective correspondence with isomorphism classes of n-pointed closed Riemann surfaces of genus g.

There is an evident forgetful map $\mathcal{T}_{g,n+1} \to \mathcal{T}_{g,n}$. This map is an analytic submersion that is equivariant over $\Gamma_g^{n+1} \to \Gamma_g^n$, and hence determines a morphism of orbifolds $\mathcal{M}_{g,n+1} \to \mathcal{M}_{g,n}$. The latter is an analytic submersion (in the orbifold sense) and the fiber over a point $p \in \mathcal{M}_{g,n}$ is the quotient of $S_{g,n}$ equipped with a complex structure defining p, modulo its (finite) group of complex automorphisms. So we might think of this morphism as the universal family of n-pointed closed Riemann surfaces of genus g.

Since $\mathcal{T}_{g,n}$ appears as a contractible universal covering of $\mathcal{M}_{g,n}$ (in the sense of orbifolds) with Galois group Γ_g^n, the rational cohomology of $\mathcal{M}_{g,n}$ is naturally isomorphic to the rational cohomology of Γ_g^n.

But $\mathcal{M}_{g,n}$ has more structure. Recall that a closed Riemann surface is in a natural way a smooth complex projective algebraic curve. So we may regard $\mathcal{M}_{g,n}$ as a moduli space of such curves. This interpretation leads to an algebraization of $\mathcal{M}_{g,n}$. Better yet: geometric invariant theory enables us to characterize $\mathcal{M}_{g,n}$ as a quasi-projective variety with the orbifold structure lifting to the structure as a stack over $\mathrm{Spec}(\mathbb{Z})$. From now on, we consider $\mathcal{M}_{g,n}$ as endowed with this structure.

5 Deligne-Mumford-Knudsen Completion

Deligne, Mumford and Knudsen [14, 55] discovered that there is a natural completion of $\mathcal{M}_{g,n}$ by allowing curves to degenerate in a mild way and that this completion has itself the interpretation of a moduli stack. The central notion here is that of *stable n-pointed curve of genus g*. This consists of a complete connected curve C of arithmetic genus g whose singularities are ordinary double points and n distinct points $x_1,\ldots,x_n$ on the smooth part of C subject to the condition that the group $\text{Aut}(C;x_1,\ldots,x_n)$ of automorphisms of C fixing these points is finite. The last condition amounts to requiring that every connected component of $C_{\text{reg}} - \{x_1,\ldots,x_n\}$ has negative Euler characteristic: no component is a smooth rational curve with at most two points removed or a smooth curve of genus one.

The local deformation theory of such curves is as nice as it could possibly be. For instance, small deformations of stable n-pointed curves of genus g are again of that type. More is true: such a curve $(C;x_1,\ldots,x_n)$ has a *universal* deformation with smooth base S of dimension $3g-3+n$. So this is given by a curve over S: $\mathcal{C} \to S$ with n disjoint sections $s_1,\ldots,s_n$ giving each fiber the structure of stable n-pointed curve of genus g, together with an identification of the closed fiber with $(C;x_1,\ldots,x_n)$. The discriminant of the morphism $\mathcal{C} \to S$ is quite simple: for every singular point p of C the locus in S parameterizing the curves where p persists as a singularity is a smooth hypersurface D_p in S and their union D is a normal crossing divisor. The group $\text{Aut}(C;x_1,\ldots x_n)$ acts naturally on the whole system.

What Deligne, Mumford and Knudsen prove is that there is a moduli stack of stable n-pointed curves of genus g, $\overline{\mathcal{M}}_{g,n}$, and that it is projective, irreducible, defined over $\text{Spec}(\mathbb{Z})$ and contains $\mathcal{M}_{g,n}$ as an open-dense subscheme. Notice that $\overline{\mathcal{M}}_{g,n}$ is locally given by a universal deformation as above. In particular, the underlying variety is at the point defined by $(C;x_1,\ldots,x_n)$ isomorphic to the quotient $\text{Aut}(C;x_1,\ldots x_n)\backslash S$. It is clear from this local picture that the Deligne-Mumford boundary $\Delta = \overline{\mathcal{M}}_{g,n} - \mathcal{M}_{g,n}$ is a normal crossing divisor (in the sense of stacks).

The generic points of this divisor parameterize n-pointed curves with a single singular point. The case where the curve is irreducible accounts for one such point; the corresponding irreducible component of Δ is usually denoted Δ_0. Otherwise the curve is a one-point union of two smooth connected projective curves, say of genera g_1 and g_2 (with $g_1 + g_2 = g$, of course) with an ensuing decomposition of $x_1,\ldots x_n$ given by a partition $I_1 \sqcup I_2$ of $\{1,\ldots,n\}$. If $g_k = 0$, then the corresponding part I_k must contain at least two elements. The set of unordered pairs $\{(g_1,I_1),(g_2,I_2)\}$, subject to this condition effectively indexes the irreducible components $\neq \Delta_0$ of the boundary divisor Δ.

The normal crossing structure of Δ defines a natural decomposition of Δ into (connected) strata. In characteristic zero a stratum parameterizes the stable pointed curves of a fixed topological type. If we remove from an n-pointed curve its singular points and the n given points, then we get a (possibly disconnected) smooth curve, hence a stratum parameterizes such curves. This can be expressed in a characteristic free manner and thus it is not difficult to see that any stratum S is a smooth stack that admits a product of moduli stacks $\prod_j \mathcal{M}_{g_j,n_j}$ as a finite cover. The closure of S is then covered by $\prod_j \overline{\mathcal{M}}_{g_j,n_j}$.

6   Covers of Moduli Stacks

The moduli stack $\overline{\mathcal{M}}_{g,n}$ admits many coverings. Any subgroup Γ of the mapping class group Γ_g^n of finite index (more precisely, a conjugacy class of those) defines a finite flat morphism $\mathcal{M}_{g,n}^\Gamma \to \mathcal{M}_{g,n}$ of stacks and then we can take the normalization $\overline{\mathcal{M}}_{g,n}^\Gamma \to \overline{\mathcal{M}}_{g,n}$ of $\overline{\mathcal{M}}_{g,n}$ in $\mathcal{M}_{g,n}^\Gamma$. In characteristic zero there is a modular interpretation of $\mathcal{M}_{g,n}^\Gamma$: it is the moduli space of smooth projective n-pointed curves $(C; x_1, \ldots, x_n)$ of genus g endowed with an isomorphism of the fundamental group of $C - \{x_1, \ldots, x_n\}$ (relative some base point) with $\pi_{g,n}$, given up an automorphism of $\pi_{g,n}$ mapping to Γ. But it is not clear whether such an interpretation is possible for its completion $\overline{\mathcal{M}}_{g,n}^\Gamma$.

Subgroups of Γ_g^n that are of particular interest are the so-called congruence subgroups. They are defined as follows: let $\pi \subset \pi_{g,n}$ be a normal subgroup of finite index that is also invariant under every automorphism that preserves the distinguished conjugacy classes $B_1, \ldots, B_n$ (see section 2). There is an evident homomorphism from Γ_g^n to the outer automorphism group of the finite group $\pi_{g,n}/\pi$. Subgroups of Γ_g^n that contain the kernel of such a homomorphism are called *congruence subgroups*. They are obviously of finite index. For the case $g = n = 1$, this yields the familiar notion of a congruence subgroup of $\Gamma_1^1 \cong \mathrm{SL}(2,\mathbb{Z})$: this is a subgroup of $\mathrm{SL}(2,\mathbb{Z})$ that contains all the matrices in $\mathrm{SL}(2,\mathbb{Z})$ congruent modulo d to the identity, for some integer d. It is well-known that there exist subgroups of finite index of $\mathrm{SL}(2,\mathbb{Z})$ that are not congruence subgroups. Ivanov has raised the question whether the situation is different for the mapping class groups Γ_g, $g \geq 2$ [47].

For some time it was not known whether for a suitable choice of Γ, the variety underlying $\overline{\mathcal{M}}_{g,n}^\Gamma$ is smooth over a given base field and this led to the foundation of an elaborate intersection theory for smooth stacks, needed to define the appropriate Chow groups. It is now known that such Γ exist and that we can take Γ to be a congruence subgroup (Looijenga [57], Pikaart-De Jong [69], Boggi-Pikaart [8]). This means that the stack $\overline{\mathcal{M}}_{g,n}$ is obtained as the orbit space of a smooth variety with respect to a finite group action. The kth Chow group of $\overline{\mathcal{M}}_{g,n}$, $\mathrm{CH}^k(\overline{\mathcal{M}}_{g,n})$, is then definable as the invariant part of the kth Chow group of this smooth variety. (Here $\mathrm{CH}^k(-)$ stands for the Chow group in codimension k defined by rational equivalence, tensorized with $\mathbb{Q}$.)

7   Tautological Classes

If $(\tilde{C}; \tilde{x}_1, \ldots, \tilde{x}_{n+1})$ is a stable $(n+1)$-pointed curve, then forgetting the last point $\tilde{x}_{n+1}$ yields a stable n-pointed curve unless the component of $\tilde{C}_{\text{reg}} - \{\tilde{x}_1, \ldots, \tilde{x}_{n+1}\}$ punctured by $\tilde{x}_{n+1}$ is a thrice punctured $\mathbb{P}^1$. But then contraction of the irreducible component of $\tilde{C}$ containing $\tilde{x}_{n+1}$ produces a stable n-pointed curve. In either case, the result is a morphism from $(\tilde{C}; \tilde{x}_1, \ldots, \tilde{x}_n)$ onto a stable n-pointed curve $(C; x_1, \ldots, x_n)$. The image x of $\tilde{x}_{n+1}$ can be any point of C and it is easy to see that the $(n+1)$-pointed curve can be recovered up to canonical isomorphism from the system $(C; x_1, \ldots, x_n; x)$.

This also works in families, so that we have a forgetful morphism $\overline{\mathcal{M}}_{g,n+1} \to \overline{\mathcal{M}}_{g,n}$ which may be thought of as the universal stable n-pointed curve of genus g. It comes in particular with n disjoint sections $s_1, \ldots, s_n$.

The functor that associates to every stable n-pointed curve its cotangent line at the ith point $(i \in \{1, \ldots, n\})$ is realized on the universal example $\overline{\mathcal{M}}_{g,n}$ as a line bundle (in the sense of stacks). This line bundle can be gotten more directly as the pull-back of the relative dualizing sheaf $\omega_{g,n}$ of the universal family $\pi : \overline{\mathcal{M}}_{g,n+1} \to \overline{\mathcal{M}}_{g,n}$ along the ith section: $s_i^* \omega_{g,n}$. The dependence on n is not entirely obvious. To see this, notice that the morphism $\tilde{C} \to C$ above defines a homomorphism $T_{x_i}^* C \to T_{\tilde{x}_i}^* \tilde{C}$. This is an isomorphism unless $\tilde{x}_i$ lies on a component that gets contracted. Which is the case precisely when $(\tilde{C}; \tilde{x}_1, \ldots, \tilde{x}_{n+1})$ defines a point on the irreducible component $\Delta_{i,n+1}$ of Δ defined by the pair $\{(g_1, I_1), (g_2, I_2)\}$ with $g_2 = 0$ and $I_2 = \{\tilde{x}_i, \tilde{x}_{n+1}\}$. So the line bundle $\tilde{s}_i^* \omega_{g,n+1}$ on $\overline{\mathcal{M}}_{g,n+1}$ is the pull-back of $s_i^* \omega_{g,n}$ on $\overline{\mathcal{M}}_{g,n}$ twisted by $\sum_{i=1}^{n} \Delta_{i,n+1}$.

We can now tell what the basic classes on $\overline{\mathcal{M}}_{g,n}$ are:

(i) the *Witten* classes

$$\psi_i := c_1(s_i^* \omega_{g,n}) \in \mathrm{CH}^1(\overline{\mathcal{M}}_{g,n}), \quad i = 1, \ldots, n,$$

(ii) the *Mumford* classes (à la Arbarello-Cornalba [1])

$$\kappa_r := \pi_!(c_1(\omega_{g,n})^{r+1}) \in \mathrm{CH}^r(\overline{\mathcal{M}}_{g,n}), \quad r = 1, 2, \ldots,$$

The *tautological subalgebra* $\mathcal{R}^\bullet(\overline{\mathcal{M}}_{g,n})$ of $\mathrm{CH}^\bullet(\overline{\mathcal{M}}_{g,n})$ is defined as follows: recall from Section 5 that the closure $\overline{S}$ of every stratum S is finitely covered by a product $\hat{S} \cong \prod_j \mathcal{M}_{g_j, n_j}$. The basic classes of the factors generate a subalgebra of $\mathrm{CH}^\bullet(\hat{S})$ whose direct image in $\mathrm{CH}^\bullet(\overline{\mathcal{M}}_{g,n})$ we denote by $A^\bullet(S)$. Then $\mathcal{R}^\bullet(\overline{\mathcal{M}}_{g,n})$ is defined as the algebra generated by the $A^\bullet(S)$, where S runs over all the strata, $\mathcal{M}_{g,n}$ included. We use the same terminology (and similar notation) for its restriction $\mathcal{R}^\bullet(\mathcal{M}_{g,n})$ to $\mathcal{M}_{g,n}$. The latter is of course already generated by the ψ_i's and the κ_r's.

The tautological algebras are respected by the obvious morphisms between moduli stacks of pointed curves, such as the pull-back and the push-forward along the finite morphisms $\hat{S} \to \overline{\mathcal{M}}_{g,n}$ (with $\hat{S}$ as above) and the projection $\overline{\mathcal{M}}_{g,n+1} \to \overline{\mathcal{M}}_{g,n}$. It is possible to characterize $\mathcal{R}^{\bullet}$ in this way as the smallest bivariant subfunctor of $CH^{\bullet}$ restricted to an appropriate category of moduli spaces of pointed curves that contains the fundamental classes of the Deligne-Mumford moduli stacks.

8 Stability

In section 4 we observed that a mapping class group and the corresponding moduli space have the same rational cohomology. So any homological property of Γ_g^n has immediate relevance for $\mathcal{M}_{g,n}$. Unfortunately, our knowledge of the homology of Γ_g^n is still rather limited. A central result is the stability theorem, due to Harer [42], that says that the homology group $H_k(\Gamma_g^n; \mathbb{Z})$ is independent of g, if g is sufficiently large (according to Ivanov, $g \geq 2k$ will do, but probably we may take $g \geq \frac{3}{2}k$). A more precise statement says how the isomorphism $H_k(\Gamma_g^n; \mathbb{Z}) \cong H_k(\Gamma_{g+1}^n; \mathbb{Z})$ is defined. There is no obvious map between the groups in question, but there is a homological correspondence defined as follows. Choose a separating circle $\delta \subset S_{g+1}$ which splits S_{g+1} into a surface of genus g and a surface S of genus one with the former containing the points labeled $x_1, \ldots, x_n$, and choose an orientation preserving homeomorphism of $S_g - \{x_{n+1}\}$ onto this component in such a way that the points $x_1, \ldots, x_n$ of S_g retain their name. If $\Gamma_{g+1,S}^n$ stands for the group of mapping classes of S_{g+1} relative to $S \cup \{x_1, \ldots, x_n\}$, then we have a natural monomorphism $\Gamma_{g+1,S}^n \to \Gamma_{g+1}^n$ and a composite epimorphism $\Gamma_{g+1,S}^n \to \Gamma_g^{n+1} \to \Gamma_g^n$. The stability theorem states that these two homomorphisms induce isomorphisms on integral homology in degree k if g is sufficiently large. We can slightly generalize the above construction to define a homomorphism

$$H_k(\Gamma_g^n; \mathbb{Z}) \otimes H_{k'}(\Gamma_{g'}^{n'}; \mathbb{Z}) \to H_{k+k'}(\Gamma_{g+g'}^{n+n'}; \mathbb{Z}),$$

provided that g and g' are sufficiently large: choose here the separating circle δ on $S_{g+g'}$ such that one piece has genus g with punctures $x_1, \ldots, x_n$ and the other has genus g' with punctures $x_{n+1}, \ldots, x_{n+m}$ and let the group of mapping classes of $S_{g+g'}$ relative to $\delta \cup \{x_1, \ldots, x_{n+n'}\}$ take the role of $\Gamma_{g+1,S}^n$.

The stable homology of the mapping class groups $\{\Gamma_g^n\}_{g=1}^{\infty}$ can be realized as the homology of a group Γ_∞^n that is defined in much the same way as Γ_g^n: replace S_g by a surface of infinite genus (but beware that such surfaces are not all mutually homeomorphic) and allow only self-homeomorphisms that are the identity outside a compact subset. In particular, we get a product $H_{\bullet}(\Gamma_\infty; \mathbb{Z}) \otimes H_{\bullet}(\Gamma_\infty; \mathbb{Z}) \to H_{\bullet}(\Gamma_\infty; \mathbb{Z})$. After tensoring with $\mathbb{Q}$, this product and the standard coproduct on $H_{\bullet}(\Gamma_\infty; \mathbb{Q})$ turn $H_{\bullet}(\Gamma_\infty; \mathbb{Q})$ into a graded-bicommutative Hopf algebra. Of course, the same applies to its dual $H^{\bullet}(\Gamma_\infty; \mathbb{Q})$.

According to a structure theorem such an algebra is as a graded algebra freely generated by its primitive subspace.

This construction can be imitated in the moduli context. Identifying the last point of an $(n+1)$-pointed smooth genus g curve with the origin of an elliptic curve produces a stable n-pointed genus $g+1$ curve. This defines a morphism $f : \mathcal{M}_{g,n+1} \times \mathcal{M}_{1,1} \to \overline{\mathcal{M}}_{g+1,n}$ whose image is an open subset of a boundary divisor. This morphism has a normal (line) bundle in the orbifold sense. Let E_f be the complement of the zero section of this normal bundle. Although there is no obvious map $E_f \to \mathcal{M}_{g+1,n}$, a tubular neighborhood theorem asserts that there is a natural homotopy class of such maps. So if we choose $p \in \mathcal{M}_{1,1}$, and let $E_f(p)$ be the restriction of E_f to $\mathcal{M}_{g,n+1} \times \{p\}$, then we have a well-defined homomorphism $H_k(E_f(p); \mathbb{Q}) \to H_k(\mathcal{M}_{g+1,n}; \mathbb{Q})$. On the other hand, projection induces a homomorphism $H_k(E_f(p); \mathbb{Q}) \to H_k(\mathcal{M}_{g,n+1}; \mathbb{Q}) \to H_k(\mathcal{M}_{g,n}; \mathbb{Q})$. These two homomorphisms are the geometric incarnations of the stability maps and hence they are isomorphisms in the stable range. In a similar fashion we get a natural homomorphism $H_k(\mathcal{M}_{g,n}; \mathbb{Q}) \otimes H_{k'}(\mathcal{M}_{g',n'}; \mathbb{Q}) \to H_{k+k'}(\mathcal{M}_{g+g',n+n'}; \mathbb{Q})$ (g and g' sufficiently large). An important feature of these homomorphisms is that they are in a sense 'motivic': they respect all the extra structure that homology groups of algebraic varieties carry, such as a mixed Hodge structure. In particular, it follows that the stable cohomology $H^\bullet(\Gamma^n_\infty; \mathbb{Q})$ has a natural mixed Hodge structure that is preserved by the coproduct (which is dual to the product defined above). It was shown by Pikaart [68] that this mixed Hodge structure is actually not mixed at all: $H^k(\Gamma^n_\infty; \mathbb{Q})$ is pure of weight k.

The tautological class κ_r introduced in 7 is, when regarded as an element of $H^{2r}(\mathcal{M}_g; \mathbb{Q})$, stable for g sufficiently large. It is not hard to prove that the corresponding element of the stable cohomology Hopf algebra is primitive. Miller [60] and Morita [61] have shown that it is nonzero and so $H^\bullet(\Gamma_\infty; \mathbb{Q})$ contains the polynomial algebra $\mathbb{Q}[\kappa_1, \kappa_2, \kappa_3, \dots]$. Mumford wrote in [64] that it seems reasonable to guess that $H^\bullet(\Gamma_\infty; \mathbb{Q})$ is no bigger than this; this 'reasonable guess' now goes under the name of Mumford's conjecture.

9 A Proarithmetic Hull of the Mapping Class Group

The lower central series of a group π is defined inductively by $\pi^{(0)} = \pi$ and $\pi^{(k+1)} = (\pi, \pi^{(k)})$. So $\pi/\pi^{(k+1)}$ is a nilpotent group. We take $\pi = \pi_g$ and note that the mapping class group Γ_g acts in $\pi_g/\pi_g^{(k+1)}$. If $\Gamma_g(k)$ denotes the image of this action, then it is clear that $\Gamma_g(0) = \mathrm{Sp}(V_g)$. It is not hard to see that $\Gamma_g(k+1)$ is an extension of $\Gamma_g(k)$ by a lattice. For $k = 0$, this lattice turns out to be just $\wedge^3_o V_g$, and as one may expect, the resulting map $T_g \to \wedge^3_o V_g$ is just the Johnson homomorphism. For higher values k, these lattices are not so easy to describe, but the least one can say is that they are obtained in a functorial manner from the symplectic lattice V_g. Things simplify a great deal if we tensor the lattice with $\mathbb{C}$: then it turns out that the resulting vector space is obtained in a functorial manner from the symplectic vector space $V_g \otimes \mathbb{C}$ (a fact that is not obvious a priori). In particular, the $\mathrm{Sp}(V_g)$ action on this vector space is algebraic in the sense that it extends to an action of the algebraic group $\mathrm{Sp}(V_g)(\mathbb{C}) = \mathrm{Sp}(V_g \otimes \mathbb{C})$. With induction one can now construct a sequence of extensions of algebraic groups by vector groups which contains $\Gamma_g(0) \leftarrow \Gamma_g(1) \leftarrow \Gamma_g(2) \leftarrow \ldots$ as a sequence of arithmetic groups. We now form the 'proarithmetic hull' of Γ_g, $\Gamma_g \to \Gamma_g(\infty) := \lim_k \Gamma_g(k)$. This map is injective, so that we may regard this as a kind of arithmetic completion of Γ_g. We are interested in the induced map on rational cohomology $H^\bullet(\Gamma_g(\infty); \mathbb{Q}) \to H^\bullet(\Gamma_g; \mathbb{Q})$. Results of Borel imply that the rational cohomology group $H^k(\Gamma_g(\infty); \mathbb{Q})$ stabilizes as $g \to \infty$ in a way that is compatible with the stabilization maps for $H^k(\Gamma_g; \mathbb{Q})$. In particular, for g sufficiently large, the image of $H^k(\Gamma_g(\infty); \mathbb{Q}) \to H^k(\Gamma_g; \mathbb{Q})$ is independent of g. Kawazumi-Morita [50] and Hain-Looijenga [40] proved that this stable image is precisely the tautological subalgebra $\mathbb{Q}[\kappa_1, \kappa_2, \kappa_3, \ldots]$. (The stable cohomology $\lim_{g \to \infty} H^\bullet(\Gamma_g(\infty); \mathbb{Q})$ is however much bigger.) This indicates that a construction of a stable class not in this algebra must be rather sophisticated. It would be interesting to see whether a similar result holds if the central lower series of π_g is replaced by the direct system of its finite index subgroups. (This completes Γ_g by the system of its congruence subgroups; the result is an 'adelization' of Γ_g.)

10   The Witten Conjecture

Given a positive integer n and an n-tuple of nonnegative integers $(k_1, k_2, \ldots, k_n)$, then for every genus g we can form the integral

$$\int_{\overline{\mathcal{M}}_{g,n}} \psi_1^{k_1} \cdots \psi_n^{k_n}.$$

This is of course zero unless $\sum_i k_i = 3g - 3 + n$ and if that equality is satisfied, we can regard it as an intersection number of tautological classes. Such a number need not be integral though, because $\overline{\mathcal{M}}_{g,n}$ is not smooth. For instance, $\int_{\overline{\mathcal{M}}_{1,1}} \psi_1 = \frac{1}{24}$. Witten [73] stated in 1989 a conjecture that predicted their values. He phrased his conjecture in terms of a generating function. In this context, the basic classes are the multiples $(2k + 1)!!\psi_i^k$ (where the double factorial stands for the product of the odd positive integers $\leq$ its argument) and therefore we find it convenient to introduce the *Witten numbers*

$$[\tau_{k_1} \cdots \tau_{k_n}]_g := (2k_1 + 1)!! \cdots (2k_n + 1)!! \int_{\overline{\mathcal{M}}_{g,n}} \psi_1^{k_1} \cdots \psi_n^{k_n}.$$

It is clear that this number is invariant under permutation of the indices. The suffix g is redundant, in the sense that the number can be nonzero for only one possible value of g. But we keep it so that we can define

$$F_g := \sum_n \frac{1}{n!} \sum_{k_1,\dots,k_n} [\tau_{k_1}\cdots\tau_{k_n}]_g t_{k_1} t_{k_2}\cdots t_{k_n} \in \mathbb{Q}[[t_0,t_1,t_2,\dots]].$$

This is a symmetric function in its variables. Note that if we give t_i degree $i-1$, then F_g is homogeneous of degree $3(g-1)$. We shall not state the original conjecture (that says that $Z := \exp(\sum_g F_g) \in \mathbb{Q}[[t_0,t_1,t_2,\dots]]$ satisfies a certain KdV-hierarchy), but give an equivalent conjecture, due to Dijkgraaf-Verlinde-Verlinde, instead. It says that Z satisfies a certain system of differential equations (known as the *Virasoro relations*). In terms of the individual F_g's these amount to:

$$\frac{\partial F_g}{\partial t_0} = \sum_{m \geq 1}(2m+1)t_m \frac{\partial F_g}{\partial t_{m-1}} + \tfrac{1}{2}\delta_{0,g}t_0^2, \qquad (\ell_{-1})$$

$$\frac{\partial F_g}{\partial t_1} = \sum_{m \geq 0}(2m+1)t_m \frac{\partial F_g}{\partial t_m} + \tfrac{1}{8}\delta_{1,g}, \qquad (\ell_0)$$

$$\frac{\partial F_g}{\partial t_{k+1}} = \sum_{m \geq 0}(2m+1)t_m \frac{\partial F_g}{\partial t_{m+k}} + \tfrac{1}{2} \sum_{m'+m''=k-1} \left(\frac{\partial^2 F_{g-1}}{\partial t_{m'}\partial t_{m''}} \right. \qquad (\ell_{k \geq 1})$$

$$\left. + \sum_{g'+g''=g} \frac{\partial F_{g'}}{\partial t_{m'}}\frac{\partial F_{g''}}{\partial t_{m''}} \right).$$

The last term of (ℓ_{-1}) resp. (ℓ_0) comes from $[\tau_0^3]_0 = 1$ resp. $[\tau_1]_1 = \frac{1}{8}$. By comparing coefficients we obtain a set of relations among the Witten numbers that allows us to calculate them recursively: equation (ℓ_k) gives $[\tau_{k_1}\cdots\tau_{k_n}\tau_{k+1}]_g$ in terms of Witten numbers involving smaller (g,n) (for the lexicographical ordering). Notice that the first two equations involve each F_g alone. They give $[\tau_{k_1}\cdots\tau_{k_n}\tau_0]_g$ and $[\tau_{k_1}\cdots\tau_{k_n}\tau_1]_g$ in terms of Witten numbers on $\overline{\mathcal{M}}_{g,n}$. These relations can be easily accounted for by means of simple intersection calculus. For $k \geq 1$, equation (ℓ_k) expresses $[\tau_{k_1}\cdots\tau_{k_n}\tau_{k+1}]_g$ in terms of Witten numbers of $\overline{\mathcal{M}}_{g,n}$ and its boundary divisors (i.e., of $\overline{\mathcal{M}}_{g-1,n+2}$ and $\overline{\mathcal{M}}_{g',n'+1} \times \overline{\mathcal{M}}_{g'',n''+1}$ with $g'+g''=g$ and $n'+n''=n$). These equations have been proved by Kontsevich [53], using a combinatorial substitute for the varieties $\overline{\mathcal{M}}_{g,n}$. It is desirable to find a purely algebro-geometric proof of these identities, because such a proof has a fair chance of generalizing to Gromov-Witten invariants (unlike the combinatorial approach).

11 Complete Subvarieties of Moduli Spaces

The moduli spaces $\mathcal{M}_{g,n}$ are not projective, with the exception of the point $\mathcal{M}_{0,3}$. This seems intuitively clear; one can deduce it from the fact that the boundary $\partial \mathcal{M}_{g,n} = \overline{\mathcal{M}}_{g,n} - \mathcal{M}_{g,n}$ in the Deligne-Mumford compactification is non-empty.

It is clear that the moduli spaces $\mathcal{M}_{0,n}$ and $\mathcal{M}_{1,n}$ are all affine. To see this in a uniform way, note that the ample divisor $\kappa_1 = 12\lambda_1 - \delta + \psi$ can be written as a sum of boundary divisors in these cases, cf. [9, 73].

Let therefore $g \geq 2$; we first consider the case $n = 0$. It is well-known that $\mathcal{M}_2$ is affine; more generally, the moduli spaces $\mathcal{H}_g$ of hyperelliptic curves of genus g are affine. In characteristic $\neq 2$, we can see this by writing a hyperelliptic curve as a double cover of $\mathbb{P}^1$ branched in $2g + 2$ distinct points; a description of $\mathcal{H}_g$ as the quotient of the complement of a hypersurface in $\mathbb{A}^{2g-1}$ by the action of the symmetric group $\mathbb{S}_{2g+2}$ results. In characteristic 2 one obtains the result starting with the observation that every hyperelliptic curve of genus g can be written in the form

$$y^2 + (1 + a_1 x + a_2 x^2 + \cdots + a_g x^g)y = x^{2g+1} + b_{g-1}x^{2g-1} + b_{g-2}x^{2g-3} + \cdots + b_2 x^5 + b_1 x^3.$$

Igusa [46] has given a description of $\mathcal{M}_2$ over $\mathbb{Z}$; in particular, in characteristics $\neq 2,5$ it is the quotient of $\mathbb{A}^3$ by a diagonal action of $\mathbb{Z}/5\mathbb{Z}$ with a unique fixed point.

For all $g \geq 3$, the moduli space $\mathcal{M}_g$ is not affine; a well-known consequence of the existence of the Satake compactification of $\mathcal{M}_g$ in which the boundary has codimension two. In particular there exist complete curves passing through any finite number of points of $\mathcal{M}_g$.

Such complete curves are not explicit, however. So for some time the problem of constructing explicit complete curves was studied. A nice solution to this problem was found by González-Díez and Harvey [38]. For all $g \geq 4$, they construct explicit complete curves in $\mathcal{M}_g$ in the following way. Take a genus 2 curve C mapping onto an elliptic curve E. Let a be a point of E different from the origin. The inverse image in $C \times C$ of the translated diagonal $\Delta_a = \{(e, e+a) : e \in E\}$ is a complete curve of pairs of distinct points. By going over to a finite cover of this curve, we obtain a complete curve of pairs of distinct points of C together with a square root of the corresponding divisor class of degree 2. That determines a complete curve of double covers of C ramified in two distinct points: a complete one-dimensional family of smooth curves of genus 4. One checks that these curves vary in moduli and obtains a complete curve in $\mathcal{M}_4$. Similarly, one finds complete curves in $\mathcal{M}_g$: start with a translate of the diagonal embedding of E in E^{2g-6} that avoids all diagonals, take its inverse image in C^{2g-6}, and form a family of double covers of C ramified in $2g - 6$ points.

In genus 3, this construction doesn't work. The genus 3 problem was solved by Zaal. Starting with a complete family of curves of genus 4 with a nonzero point of order two in the Jacobian, one obtains a complete family of 3-dimensional Prym varieties. Zaal showed that suitable choices guarantee that all these Pryms are Jacobians of smooth curves [74]. (For a very different solution, see part II of [38].)

What about complete subvarieties of $\mathcal{M}_g$ of higher dimension? We cannot use the Satake compactification (it appears); almost all results rely on a variant of the classical Kodaira construction. Kodaira observed that one may construct an explicit complete curve in $\mathcal{M}_6$ by starting with a genus 3 curve C that is a double unramified cover of a genus 2 curve. This gives a complete curve of pairs of distinct points of C; one proceeds as above and obtains the result. The construction can be repeated, since one finds in fact a complete one-dimensional family of curves of genus 6 with a pair of distinct points, the ramification points coming from the double cover of C. The monodromy problems arising in the choice of a square root can always be resolved, so this leads to a complete surface in $\mathcal{M}_{12}$, a complete threefold in $\mathcal{M}_{24}$, etc. This can be improved upon by starting with the complete curve in $\mathcal{M}_4$ of [38]: one finds a complete surface in $\mathcal{M}_8$, a complete threefold in $\mathcal{M}_{16}$, etc. Another variant is to use triple covers ramified in one point: if the covered curve has genus h, the cover has genus $3h - 1$. Asymptotically this doesn't lead to better results, but since one can start with a complete curve in $\mathcal{M}_3$, one obtains another construction of a complete surface in $\mathcal{M}_8$.

A new development occurs here through the recent work of Zaal [75]. Using the important work of Keel [51], Zaal constructs in characteristic $p > 0$ a complete surface in $\mathcal{M}_{3,2}$. (At the moment it is not clear yet whether it is also possible to do this in characteristic 0.) This leads to a complete surface in $\mathcal{M}_6$ in characteristic $p > 2$ via double covers. One also finds a complete threefold in $\mathcal{M}_{12}$, etc.

Observe that all known complete subvarieties of $\mathcal{M}_g$ of dimension > 1 lie in the locus of curves that admit a map onto a curve of lower (but positive) genus. (Every component of this locus has codimension $\geq g - 1$.) In particular, we don't know whether a complete surface in $\mathcal{M}_g$ could contain a general point. Perhaps it is more important to study complete subvarieties passing through a general point than arbitrary ones, cf. [45], p. 55. In [65] Nicorestianu shows that the base of a complete, generically non-degenerate 2-dimensional family of smooth curves (of genus $g \geq 4$) is necessarily a surface of general type (in characteristic 0).

A celebrated result is Diaz's upper bound $g - 2$ for the dimension of a complete subvariety of $\mathcal{M}_g$, see [15]. Looijenga's result on the tautological ring of $\mathcal{M}_g$ gives a different proof, valid also in positive characteristic [58]. Note that this bound is known to be sharp only for $g = 2$ and 3; since we don't know whether $\mathcal{M}_4$ contains a complete surface, it might be argued that we don't understand curves of genus 4.

The class $\lambda_g \lambda_{g-1}$ in the Chow ring of $\overline{\mathcal{M}}_g$ vanishes on the boundary $\overline{\mathcal{M}}_g - \mathcal{M}_g$ ([23], see also [24]). Therefore it (or a positive multiple of it) is a candidate for the class of a complete subvariety of $\mathcal{M}_g$ of dimension $g - 2$, if that exists. One might also phrase the existence of this class as the absence of an intersection-theoretical obstruction for the existence of a complete subvariety of dimension $g - 2$. Compare the discussion in [40], §5. In genus 4 the class $\lambda_4 \lambda_3$ probably is the only candidate *in cohomology* for the class of a complete surface. This would follow from the calculation of the codimension 2 Chow group of $\overline{\mathcal{M}}_4$ [22] if $H^4(\overline{\mathcal{M}}_4)$ is generated by tautological classes (cf. [2], discussed in section 12, and [18], where Edidin shows that $H^4(\overline{\mathcal{M}}_g)$ is generated by tautological classes in the stable range). In [24] it is pointed out that the structure of the tautological ring of $\mathcal{M}_g$ (known for $g \leq 15$) suggests that there are no constraints on complete subvarieties of dimension $\leq g/3$, while there are many constraints on complete subvarieties of dimension $g - 2$; so that it might be a better idea to look for the former rather than the latter. Zaal's construction of complete surfaces in $\mathcal{M}_6$ in characteristic $p > 2$ might be considered as evidence for this idea. (On p. 57 of

[45] it is stated that the maximal dimension of a complete subvariety of $\mathcal{M}_6$ (over $\mathbb{C}$) is known to be at least 2, but this appears to be a typo.)

Diaz's original motivation [16] for finding an upper bound for the dimension of a complete subvariety of $\mathcal{M}_g$ was the implication that a family of curves whose image in moduli has larger dimension, necessarily degenerates—for many types of questions, this knowledge can be of great help. In the same spirit, he shows that a complete subvariety of $\overline{\mathcal{M}}_g$ of dimension $\geq 2g - 2$ necessarily meets Δ_0, the divisor of irreducible singular curves and their degenerations ([16], p. 80, Corollary; $2g - 2$ is certainly what is intended). In other words, one has the upper bound $2g - 3$ for the dimension of a complete subvariety of the moduli space $\widetilde{\mathcal{M}}_g = \overline{\mathcal{M}}_g - \Delta_0$ of curves of compact type. This bound is a direct corollary of the bound for $\mathcal{M}_g$, hence it holds in all characteristics as well. The surprise is that in positive characteristic the bound $2g - 3$ for $\widetilde{\mathcal{M}}_g$ is known to be sharp. One obtains this result from a consideration of the locus $V_0(\overline{\mathcal{M}}_g) = V_0(\widetilde{\mathcal{M}}_g)$ of stable curves with p-rank 0. In [26] it is shown that it is pure of codimension g. The role that the class $\lambda_g \lambda_{g-1}$ played in relation to $\mathcal{M}_g$ is now played by λ_g: it vanishes on Δ_0 and has the right codimension. Van der Geer [28] (this volume) explicitly determined the class of $V_0(\mathcal{A}_g)$; it is a multiple of λ_g, hence the same holds for the class of $V_0(\widetilde{\mathcal{M}}_g)$.

The fact that the bound $2g - 3$ for $\widetilde{\mathcal{M}}_g$ is sharp in positive characteristic (more precisely, that the known maximal complete subvariety of $\widetilde{\mathcal{M}}_g$ occurs only in characteristic $p > 0$) as well as Keel's result [51] that the relative dualizing sheaf of $\overline{\mathcal{M}}_{g,1}$ over $\overline{\mathcal{M}}_g$ is semi-ample in positive characteristic, but not in characteristic 0, lead to the idea that the maximal dimension of a complete subvariety of $\mathcal{M}_g$ or of $\widetilde{\mathcal{M}}_g$ or of $\mathcal{A}_g$ may well depend on the characteristic. This is poignantly expressed by a conjecture of Oort (conjecture 2.3 G in [66]) that $\mathcal{A}_3$ over $\mathbb{C}$ does *not* contain a complete threefold.

Equivalently, $\widetilde{\mathcal{M}}_3$ over $\mathbb{C}$ would not contain a complete threefold. Even the answer to the following question appears to be unknown (cf. [65] for $g = 3$):

Question 11.1. Does the moduli space $\widetilde{\mathcal{H}}_g \otimes \mathbb{C}$ of complex hyperelliptic curves of compact type of genus $g \geq 3$ contain a complete surface?

It is easy to see that it contains a complete curve. The existence of a complete surface in $\widetilde{\mathcal{H}}_3 \otimes \mathbb{C}$ is a necessary condition for the existence of a complete threefold in $\widetilde{\mathcal{M}}_3 \otimes \mathbb{C}$. The question can be formulated in terms of genus 0 curves, so it should be more approachable.

Finally a brief discussion of $\mathcal{M}_{g,n}$ in case $n > 0$ (and $g \geq 2$). There are obvious relations to the case $n = 0$, for different values of the genus. As mentioned above, $\mathcal{M}_{3,2}$ contains a complete surface in positive characteristic [75], while this is not known in characteristic 0. (The construction of the complete surface works for all $g \geq 3$.) Since the projection $\mathcal{M}_{g,1} \to \mathcal{M}_g$ is projective, while the projections $\mathcal{M}_{g,n} \to \mathcal{M}_{g,1}$ are affine, the Diaz-bound for $\mathcal{M}_{g,1}$ is $g-1$, while for $n > 1$ it is at most $g-1$. The only existence result we know is that $\mathcal{M}_{g,n}$ is never affine for $n > 0$ and $g \geq 2$. For $n > 1$ there are fibers of $\mathcal{M}_{g,n}$ over $\mathcal{M}_g$ that contain a complete curve: take a curve C of genus g mapping onto an elliptic curve E, and proceed as in [38], discussed above. Note also that $C \times C - \Delta$ always contains complete curves: the difference map $(p,q) \mapsto p - q$ to the surface $C - C$ in the jacobian contracts the diagonal (Van Geemen). The following question seems relevant:

Question 11.2. For which smooth curves C of genus $g \geq 2$ does the complement in $C \times C \times C$ of all diagonals contain a complete curve?

We end with a conjecture:

Conjecture 11.3. *(Looijenga)* $\mathcal{M}_g$ *can be covered with* $g - 1$ *affine opens.*

Harer's bound $4g - 5$ for the cohomological dimension of $\mathcal{M}_g$ [43] would be one of several consequences of this result.

12   Intersection Theory

Here we discuss some of the developments regarding the Chow, tautological, and cohomology rings of the moduli spaces $\mathcal{M}_{g,n}$ and $\overline{\mathcal{M}}_{g,n}$ since the survey [40] was written. Those directly related to Gromov-Witten theory will be reviewed in section 13.

A great deal of progress has been made in genus 1. Getzler [30, 31, 32] has calculated the $\mathbb{S}_n$-equivariant Serre polynomials of $\mathcal{M}_{1,n}$ and $\overline{\mathcal{M}}_{1,n}$. Hence the $\mathbb{S}_n$-representations $H^{p,q}(\overline{\mathcal{M}}_{1,n})$ are known. In particular, $H^{0,11}(\overline{\mathcal{M}}_{1,11})$ is one-dimensional, which was known before via Eichler-Shimura theory, cf. [13, 70, 71]; the representation is the alternating one. (The corresponding 2-dimensional Hodge structure of weight 11 is associated to the discriminant cusp form Δ.) It follows that $\overline{\mathcal{M}}_{1,n}$ is not unirational for all $n \geq 11$. With a beautiful construction, Belorousski [5] has shown that $\overline{\mathcal{M}}_{1,10}$ is rational, so that $\overline{\mathcal{M}}_{1,n}$ is unirational for all $n \leq 10$. In fact, using analogous constructions he shows that the Chow ring of $\overline{\mathcal{M}}_{1,n}$ (with $\mathbb{Q}$-coefficients as always) is generated by boundary cycles for $n \leq 10$. By induction this implies that the Chow ring of $\overline{\mathcal{M}}_{1,n}$ equals the tautological ring for $n \leq 10$. This cannot hold (over $\mathbb{C}$) for any $n \geq 11$ by (a suitable extension of) Jannsen's result ([48], Thm. 3.6, Rem. 3.11).

A crucial case is $n = 4$. Getzler's calculation implies that the $\mathbb{S}_4$-invariant part of $H^4(\overline{\mathcal{M}}_{1,4})$ is 7-dimensional. But there are 9 invariant boundary cycles, with only one WDVV-relation (i.e., coming from $\overline{\mathcal{M}}_{0,4}$) between them. Hence there is here a new, genus 1, relation. Getzler computes it in [33]. (He also announces there a proof that the even-dimensional homology of $\overline{\mathcal{M}}_{1,n}$ is spanned by boundary cycles, and that all relations among these cycles come from this genus 1 relation and the genus 0 relations.) In [67], Pandharipande uses the Hurwitz scheme to construct Getzler's relation algebraically, and Belorousski uses this to analyze the Chow rings of $\overline{\mathcal{M}}_{1,n}$ for low n in detail. E.g., he shows that the tautological ring (or equivalently, the ring generated by boundary cycles) is multiplicatively generated by divisors for $n \leq 5$, while for $n \geq 6$ it is generated in codimensions one and two.

For $n \leq 5$, he also obtains explicit presentations of $A^\bullet(\overline{\mathcal{M}}_{1,n})$. Returning to $n = 4$, by identifying the 4 points in 2 pairs, one obtains a map $\overline{\mathcal{M}}_{1,4} \to \overline{\mathcal{M}}_3$. Getzler's relation pushes forward to a relation in $A^4(\overline{\mathcal{M}}_3)$. Belorousski and Pandharipande verified that the obtained relation equals (modulo genus 0 relations) the non-trivial relation found in [21], Lemma 4.4, from associativity(!) considerations.[1]

In genus 2, far less is known. Mumford [64] determined the Chow ring of $\overline{\mathcal{M}}_2$, and it is not hard to determine $A^\bullet(\overline{\mathcal{M}}_{2,1})$ from his results. The Chow, tautological, and cohomology rings coincide here. With a delicate calculation using lots of ingredients, Getzler [34] computes the $\mathbb{S}_n$-equivariant Serre polynomials of $\overline{\mathcal{M}}_{2,n}$ for $n = 2$ and 3. He also computes the cohomology ring $H^\bullet(\overline{\mathcal{M}}_{2,2})$ and announces the result for $n = 3$. In particular, $h^4(\overline{\mathcal{M}}_{2,3}) = 44$. This result is the starting point for [6]. As Belorousski and Pandharipande point out, there are 47 *descendent stratum classes* in $A^2(\overline{\mathcal{M}}_{2,3})$. (It is not hard to see that Getzler's topological recursion relations [34] are algebraic, so these 47 classes span $\mathcal{R}^2(\overline{\mathcal{M}}_{2,3})$.) Exactly 2 relations come from genus 0, none from genus 1, so there must exist a new, genus 2 relation in homology. Belorousski and Pandharipande construct such a relation algebraically using admissible double covers. Bini, Gaiffi and Polito [7] have computed the generating function for the Euler characteristic of $\overline{\mathcal{M}}_{2,n}$.

If one believes the conjecture [24] (this volume) that the tautological ring $\mathcal{R}^\bullet(\mathcal{M}_g)$ satisfies Poincaré duality, then it is quite reasonable to believe that the same holds for $\mathcal{R}^\bullet(\overline{\mathcal{M}}_g)$, especially because of the role that the classes $\lambda_g \lambda_{g-1}$ resp. λ_g play in these cases (cf. §11 and [40]). These classes also tell us what in this respect the correct moduli spaces of pointed curves should be: $\widehat{\mathcal{M}}_{g,n} = \pi^{-1}(\mathcal{M}_g) \subset \overline{\mathcal{M}}_{g,n}$ resp. $\widetilde{\mathcal{M}}_{g,n}$. The first thing to observe is that the tautological rings of these spaces are one-dimensional in codimension $g - 2 + n$ resp. $2g - 3 + n$ and vanish in higher codimensions (this follows quite easily from [58] and [23], cf. [40], p. 108). The assumption that the tautological rings of these moduli spaces satisfy Poincaré duality can be used to predict (but not prove) relations of the type we saw above for $\overline{\mathcal{M}}_{1,4}$ and $\overline{\mathcal{M}}_{2,3}$. (Both for $\widetilde{\mathcal{M}}_{1,4} = \widehat{\mathcal{M}}_{1,4}$ and for $\widehat{\mathcal{M}}_{2,3}$ the $\mathbb{S}_n$-invariant part of $\mathcal{R}^1$ is 3-dimensional, while there are 4 invariant generators in degree 2, which is assumed to be dual to degree 1.)

In their recent paper [2], Arbarello and Cornalba show how one can in principle compute the low degree cohomology groups $H^k(\overline{\mathcal{M}}_{g,n})$ for k fixed and arbitrary g and n. Their elegant method proceeds as follows. If the boundary divisor $\partial \mathcal{M}_{g,n}$ were ample, $H^k(\overline{\mathcal{M}}_{g,n})$ would inject for low k into $H^k(\partial \mathcal{M}_{g,n})$. It hardly ever is ample, but Harer's calculation [43] of the virtual cohomological dimension of $\mathcal{M}_{g,n}$ implies that $H^k(\overline{\mathcal{M}}_{g,n}) \to H^k(\partial \mathcal{M}_{g,n})$ is injective for $k \leq d(g,n)$ (where $d(g,n) = 2g - 3 + n$ for $g,n > 0$, while $d(g,0) = d(g,1) = 2g - 2$ and $d(0,n) = n - 4$). Mixed Hodge theory shows that the map $H^k(\overline{\mathcal{M}}_{g,n}) \to H^k(N)$

---

[1] While doing the calculation, they discovered that some of the genus 0 relations in [21] are stated incorrectly. The correct relations are $[(7)] = 3[(6)]$ and $\delta_1[(e)]_Q = [(6)]_Q = \frac{2}{3}[(7)]_Q$ in codimension 4 and $[(c)] + [(e)] = 2[(d)]$, $[(c)] = 3[(b)]$, $\delta_1[(2)]_Q = [(b)]_Q = \frac{4}{3}[(c)]_Q = \delta_0[(6)]_Q$, $\lambda[(6)]_Q = \frac{1}{9}[(c)]_Q$, $\delta_1[(6)]_Q = -\frac{1}{9}[(c)]_Q - \frac{2}{3}[(f)]_Q$ in codimension 5 (cf. Thm. 3.1, p. 385, p. 400, Lemma 4.5, p. 403, Table 8).

is then injective as well, where N is the normalization of $\partial \mathcal{M}_{g,n}$. Now one uses the structure of N and a double induction on g and n to compute some of the low degree cohomology groups. E.g., for odd k, if one shows $H^k(\overline{\mathcal{M}}_{g,n}) = 0$ for all (g,n) with $d(g,n) < k$, then $H^k(\overline{\mathcal{M}}_{g,n}) = 0$ for all (g,n). So the proof that $H^1(\overline{\mathcal{M}}_{g,n}) = 0$ is reduced to checking it for the point $\overline{\mathcal{M}}_{0,3}$ and the projective lines $\overline{\mathcal{M}}_{0,4}$ and $\overline{\mathcal{M}}_{1,1}$! By checking more *seed cases*, Arbarello and Cornalba prove that $H^3(\overline{\mathcal{M}}_{g,n})$ and $H^5(\overline{\mathcal{M}}_{g,n})$ vanish for all (g,n) (this uses results of Getzler and Looijenga).

For low even k, one would like to show that $H^k(\overline{\mathcal{M}}_{g,n})$ is generated by tautological classes. At present this is known for $k = 2$. Arguing by induction, one assumes that $H^2(\overline{\mathcal{M}}_{h,m})$ is tautological for the moduli spaces $\overline{\mathcal{M}}_{h,m}$ appearing in $\partial \mathcal{M}_{g,n}$. Writing $N = \coprod_i X_i$, one knows that $f = \oplus_i f_i : H^2(\overline{\mathcal{M}}_{g,n}) \to \oplus_i H^2(X_i)$ is injective when $d(g,n) \geq 2$. Now on the one hand one knows exactly what happens to the tautological classes under f, which provides a lower bound for $\mathrm{Im}(f)$. But on the other hand, any collection of classes $(f_i(\alpha))_i \in \oplus_i H^2(X_i)$ satisfies obvious compatibility relations on the "intersections" of the X_i. Since by the induction hypothesis the $H^2(X_i)$ are tautological, the upper bound for $\mathrm{Im}(f)$ that this gives can be described exactly. The beautiful idea is that the lower and upper bound coincide, essentially. The low genus cases have to be treated carefully because the tautological classes are not independent. In this way one obtains a different proof of Harer's result [41] that $H^2(\overline{\mathcal{M}}_{g,n})$ is tautological.

13 Stable Maps and the Virasoro Conjecture

A general method to define invariants of a space X, that was developed through the work of Donaldson, Gromov, Witten, Kontsevich, and many others, is to consider an auxiliary space, e.g., a space of maps of curves to X, and then to compute a (well-defined) "natural" integral on that auxiliary space. In algebraic geometry, a breakthrough occurred through Kontsevich's construction of the space of stable maps. In this section, we briefly review the basic definitions and formulate some of the most important results. Then we discuss the Virasoro conjecture. We will show how this theory has repercussions for the study of $\overline{\mathcal{M}}_{g,n}$ itself—somewhat contrary to its original motivation.

Let X be a nonsingular complex projective variety, and let β be a class in $H_2(X,\mathbb{Z})$. One can consider the moduli stack $\mathcal{M}_{g,n}(X,\beta)$ classifying n-pointed smooth curves of genus g with a map $f : C \to X$ satisfying $f_*([C]) = \beta$. The *expected dimension* of this stack is

$$3g - 3 + n + \chi(f^*T_X) = 3g - 3 + n + (\dim X)(1 - g) - K_X \cdot \beta.$$

One seeks to compactify this space in a natural way; as Kontsevich [54] showed, this can be done using *stable maps*. A map $f : C \to X$ from a reduced, connected, nodal, n-pointed curve C of genus g to X, with $f_*([C]) = \beta$, is stable if each nonsingular rational component of C that is mapped to a point contains at least three special (nodal or marked) points, and each component of genus 1 that is mapped to a point contains at least one special point. (Equivalently, the map has finitely many automorphisms.)

Fulton and Pandharipande [27] explain in detail how a projective coarse moduli space $\overline{\mathcal{M}}_{g,n}(X,\beta)$ of stable maps can be constructed. When X is a point (hence $\beta = 0$), one recovers the Deligne-Mumford-Knudsen moduli space $\overline{\mathcal{M}}_{g,n}$ of stable n-pointed curves of genus g. In general, however, $\overline{\mathcal{M}}_{g,n}(X,\beta)$ is reducible, singular, nonreduced, and has components whose dimension is not the expected one. That one can nevertheless do intersection theory on this space is something of a miracle; it is possible thanks to the construction of the virtual fundamental class [56, 4, 3]. This cycle $[\overline{\mathcal{M}}_{g,n}(X,\beta)]^{vir}$ lives in the expected dimension and satisfies the axioms of Gromov-Witten theory given by Kontsevich and Manin.

Natural cohomology classes on $\overline{\mathcal{M}}_{g,n}(X,\beta)$ arise in two ways. Via the n evaluation morphisms $e_i : \overline{\mathcal{M}}_{g,n}(X,\beta) \to X$ sending a stable map to the image of the i-th marked point, one can pull back cohomology classes from X. (Fix an additive homogeneous basis $1 = T_0, T_1, \ldots, T_m$ of $H^\bullet(X) \otimes \mathbb{Q}$.) One also has the first Chern classes of the n cotangent line bundles $\mathcal{L}_i = s_i^* \omega_{\mathcal{U}/\overline{\mathcal{M}}}$, where $\mathcal{U}$ is the universal curve and s_i the section corresponding to the i-th marked point. Define classes

$$\tau_k^j = \tau_k^j(i) = e_i^*(T_j) \cup c_1(\mathcal{L}_i)^k .$$

Gromov-Witten invariants, and their descendents (i.e., τ_k^j with $k > 0$ occur), are defined in the algebraic context by integrating these classes against the virtual fundamental class:

$$\langle \tau_{k_1}^{j_1} \cdots \tau_{k_n}^{j_n} \rangle_{g,n,\beta} = \langle \tau_{k_1}^{j_1}(1) \cdots \tau_{k_n}^{j_n}(n) \rangle_{g,n,\beta} := \int_{[\overline{\mathcal{M}}_{g,n}(X,\beta)]^{vir}} \prod_{i=1}^{n} \tau_{k_i}^{j_i}(i)$$

(some care is required when X has odd-dimensional cohomology classes). More generally, one has the Gromov-Witten classes (or "full system of Gromov-Witten invariants"):

$$I_{g,n}^\beta(T_{j_1}, \ldots, T_{j_n}) = \pi_*([\overline{\mathcal{M}}_{g,n}(X,\beta)]^{vir} \cap e_1^*(T_{j_1}) \cap \cdots \cap e_n^*(T_{j_n})) \in H^\bullet(\overline{\mathcal{M}}_{g,n}, \mathbb{Q})$$

with $\pi : \overline{\mathcal{M}}_{g,n}(X,\beta) \to \overline{\mathcal{M}}_{g,n}$ the forgetful map $(2g - 2 + n > 0)$.

There are several ways to produce relations between Gromov-Witten invariants:

a. Under the natural map $\pi : \overline{\mathcal{M}}_{g,n+1}(X,\beta) \to \overline{\mathcal{M}}_{g,n}(X,\beta)$ the virtual fundamental class pulls back to the virtual fundamental class. Also, $c_1(\mathcal{L}_i) = \pi^* c_1(\mathcal{L}_i) + \Delta_{i,n+1}$, where $\Delta_{i,n+1}$ is the divisor 'where the i-th and $(n+1)$-st point have come together'. This leads to generalized string and dilaton equations, expressing GW invariants involving a τ_0^0 resp. a τ_1^0 in terms of simpler GW invariants, and a divisor equation, for GW invariants involving a τ_0^D, where D is a divisor class.

b. Relations between the classes of the strata (and their descendents) occurring
in the topological stratification of $\overline{\mathcal{M}}_{g,n}$ yield relations between GW invariants, via the 'splitting axiom'. Examples of such relations were given in
section 12. One also has the so-called topological recursion relations (TRR)
that in genus 0 and 1 express the class of a cotangent line in boundary divisor classes. In higher genus, this is not possible. Getzler [34] conjectures that
monomials of degree g in cotangent line classes can be expressed in terms of
boundary classes, and proves this (explicitly) for genus 2.

c. In case X admits a torus action satisfying certain conditions, one can attempt
to compute GW invariants using Bott localization. Ellingsrud and Strømme
[20] introduced Bott's formula to enumerative geometry. Subsequently Kontsevich [54] used localization to compute GW invariants in genus 0. In higher
genus, one needs a localization formula for the virtual fundamental class;
this was accomplished by Graber and Pandharipande [39]. One should note
that the use of this method is not restricted to the calculation of GW invariants on X with a torus action: certain GW invariants can be expressed
in terms of an ambient projective space. Compare the work of Kontsevich
[54] expressing the number of rational curves on a quintic threefold as a sum
over trees—the first step in Givental's solution [37] of the mirror conjecture.

We end with the Virasoro conjecture of Eguchi, Hori and Xiong [19]. A proof
of it will lead to a wealth of relations between GW invariants and their descendents (although the extent to which it determines all such invariants is not clear
at the moment). Just as in section 10, one organizes the GW invariants and their
descendents for a fixed X into a generating function, the so-called full gravitational potential function. Eguchi, Hori and Xiong conjecture that its exponential
is annihilated by certain formal differential operators that form a representation of the affine Virasoro algebra. (The initial form of the conjecture was for X
with only (p,p) cohomology; the extension to general X is due to Katz. See, e.g.,
[11, 36].) There is considerable evidence for the Virasoro conjecture, but we will
not discuss this here (see e.g. [35]). Instead, we mention the work of Getzler and
Pandharipande [36] who investigate the implications of the Virasoro conjecture
in the case $\beta = 0$. There is a natural isomorphism $\overline{\mathcal{M}}_{g,n}(X,0) = \overline{\mathcal{M}}_{g,n} \times X$ under
which the virtual fundamental class is identified with the top Chern class of the
exterior tensor product of the dual of the Hodge bundle on $\overline{\mathcal{M}}_{g,n}$ and the tangent
bundle of X. Getzler and Pandharipande show that for $X = \mathbb{P}^2$ this case of the
Virasoro conjecture implies the conjectured proportionality formulas [24] (this
volume) for the tautological ring of $\mathcal{M}_g$, while for $X = \mathbb{P}^1$ it implies the beautiful
identities

$$\int_{\overline{\mathcal{M}}_{g,n}} \lambda_g \prod_{i=1}^{n} \psi_i^{a_i} = \binom{2g-3+n}{a_1\, a_2\, \cdots\, a_n} \int_{\overline{\mathcal{M}}_{g,1}} \lambda_g \psi_1^{2g-2}$$

for a_i with sum $2g-3+n$. In [25] the integral on the right side is computed using
virtual localization [39] and more classical techniques.

Acknowledgements

C.F. thanks the Max-Planck-Institut für Mathematik, Bonn, for excellent working conditions and support.

Bibliography

[1] E. Arbarello, M. Cornalba: *Combinatorial and algebro-geometric cohomology classes on the moduli spaces of curves.* J. Algebraic Geom. 5 (1996), no. 4, 705–749.

[2] E. Arbarello, M. Cornalba: *Calculating cohomology groups of moduli spaces of curves via algebraic geometry.* Preprint 1998, math.AG/9803001.

[3] K. Behrend, *Gromov-Witten invariants in algebraic geometry,* Invent. Math. 127 (1997), 601–617.

[4] K. Behrend and B. Fantechi, *The intrinsic normal cone,* Invent. Math. 128 (1997), 45–88.

[5] P. Belorousski: *Chow rings of moduli spaces of pointed elliptic curves.* Thesis, University of Chicago, 1998.

[6] P. Belorousski, R. Pandharipande: *A descendent relation in genus 2.* Preprint 1998, math.AG/9803072.

[7] G. Bini, G. Gaiffi, M. Polito: *A formula for the Euler characteristic of* $\overline{\mathcal{M}}_{2,n}$. Preprint 1998, math.AG/9806048.

[8] M. Boggi, M. Pikaart: Galois covers of moduli of curves, Utrecht 1997, to appear in Comp. Math.

[9] M. Cornalba: *On the projectivity of the moduli spaces of curves.* J. Reine Angew. Math. 443 (1993), 11–20.

[10] M. Cornalba, X. Gómez-Mont, A. Verjovsky, eds., *Lectures on Riemann surfaces.* World Scientific Publishing Co., Inc., Teaneck, NJ, 1989.

[11] D. Cox, S. Katz: *Mirror symmetry and algebraic geometry.* Mathematical Surveys and Monographs, 68. AMS, Providence 1999.

[12] M. Dehn: *Die Gruppe der Abbildungsklassen,* Acta Math. 69 (1938), 135–206.

[13] P. Deligne: *Formes modulaires et représentations ℓ-adiques.* Séminaire Bourbaki (1970/71), Exp. No. 355, pp. 139–172. Lecture Notes in Math., Vol. 179, Springer, Berlin, 1971.

[14] P. Deligne and D. Mumford: *The irreducibility of the space of curves of given genus,* Inst. Hautes Études Sci. Publ. Math. 36 (1969), 75–109.

[15] S. Diaz: *A bound on the dimensions of complete subvarieties of* $\mathcal{M}_g$. Duke Math. J. 51 (1984), no. 2, 405–408.

[16] S. Diaz: *Complete subvarieties of the moduli space of smooth curves.* In: *Algebraic geometry—Bowdoin 1985* (S. Bloch, ed.), 77–81, Proc. Sympos. Pure Math., 46, Part 1, AMS, Providence, 1987.

[17] R. Dijkgraaf, C. Faber, G. van der Geer, eds., *The moduli space of curves.* Progress in Mathematics 129, Birkhäuser, Boston 1995.

[18] D. Edidin: *The codimension-two homology of the moduli space of stable curves is algebraic.* Duke Math. J. 67 (1992), 241–272.

[19] T. Eguchi, K. Hori, C.-S. Xiong: *Quantum cohomology and Virasoro algebra.* Phys. Lett. B 402 (1997), 71–80.

[20] G. Ellingsrud, S.A. Strømme: *Bott's formula and enumerative geometry.* J. Amer. Math. Soc. 9 (1996), no. 1, 175–193.

[21] C. Faber: *Chow rings of moduli spaces of curves I: The Chow ring of $\overline{\mathcal{M}}_3$.* Ann. of Math. 132 (1990), 331–419.

[22] C. Faber: *Chow rings of moduli spaces of curves II: Some results on the Chow ring of $\overline{\mathcal{M}}_4$.* Ann. of Math. 132 (1990), 421–449.

[23] C. Faber: *A non-vanishing result for the tautological ring of $\mathcal{M}_g$.* Preprint 1995, math.AG/9711219.

[24] C. Faber: *A conjectural description of the tautological ring of the moduli space of curves.* This volume, 109–129.

[25] C. Faber, R. Pandharipande, *Hodge integrals and Gromov-Witten theory*, preprint 1998, math.AG/9810173.

[26] C. Faber, G. van der Geer: *Complete subvarieties of moduli spaces and the Prym map.* In preparation.

[27] W. Fulton, R. Pandharipande: *Notes on stable maps and quantum cohomology.* In *Algebraic geometry—Santa Cruz 1995* (J. Kollár, R. Lazarsfeld, D. Morrison, eds.), 45–96, Proc. Symp. Pure Math. 62, Part 2, AMS, Providence 1997.

[28] G. van der Geer: *Cycles on the moduli space of abelian varieties.* This volume, 65–89.

[29] S. Gervais: *A finite presentation of the mapping class group of an oriented surface,* preprint Univ. de Nantes (1998).

[30] E. Getzler: *Mixed Hodge structures of configuration spaces.* Preprint 1995, math.AG/9510018.

[31] E. Getzler: *Resolving mixed Hodge modules on configuration spaces.* Preprint 1996, math.AG/9611003, to appear in Duke Math. J.

[32] E. Getzler: *The semi-classical approximation for modular operads.* Comm. Math. Phys. 194 (1998), 481–492.

[33] E. Getzler: *Intersection theory on $\overline{\mathcal{M}}_{1,4}$ and elliptic Gromov-Witten invariants.* J. Amer. Math. Soc. 10 (1997), 973–998.

[34] E. Getzler: *Topological recursion relations in genus 2.* Preprint 1998, math.AG/9801003.

[35] E. Getzler: *The Virasoro conjecture for Gromov-Witten invariants.* Preprint 1998, math.AG/9812026.

[36] E. Getzler, R. Pandharipande: *Virasoro constraints and the Chern classes of the Hodge bundle.* Nucl. Phys. B 530 (1998), 701–714.

[37] A. Givental: *Equivariant Gromov-Witten invariants.* Internat. Math. Res. Notices 1996, no. 13, 613–663.

[38] G. González Díez, W.J. Harvey: *On complete curves in moduli space. I, II.* Math. Proc. Cambridge Philos. Soc. 110 (1991), no. 3, 461–466, 467–472.

[39] T. Graber, R. Pandharipande: *Localization of virtual classes.* Invent. Math. 135 (1999), 487–518.

[40] R. Hain, E. Looijenga, *Mapping class groups and moduli spaces of curves.* In *Algebraic geometry—Santa Cruz 1995* (J. Kollár, R. Lazarsfeld, D. Morrison, eds.), 97–142, Proc. Symp. Pure Math. 62, Part 2, AMS, Providence 1997.

[41] J. Harer: *The second homology group of the mapping class group of an orientable surface.* Invent. Math. 72 (1983), no. 2, 221–239.

[42] J. Harer: *Stability of the homology of the mapping class groups of orientable surfaces,* Ann. of Math. 121 (1985), 215–249.

[43] J. Harer: *The virtual cohomological dimension of the mapping class group of an orientable surface.* Invent. Math. 84 (1986), no. 1, 157–176.

[44] J. Harris: *Curves and their moduli.* In: *Algebraic geometry—Bowdoin 1985* (S. Bloch, ed.), 99–143, Proc. Sympos. Pure Math., 46, Part 1, AMS, Providence 1987.

[45] J. Harris, I. Morrison: *Moduli of curves.* GTM 187, Springer, Berlin-New York, 1998.

[46] J.I. Igusa: *Arithmetic variety of moduli for genus two.* Ann. of Math. (2) 72 1960 612–649.

[47] N. Ivanov: *Ten Problems on the Mapping Class Groups,* Preprint Michigan State University. (Available from http://www.math.msu.edu/ ivanov/).

[48] U. Jannsen: *Motivic sheaves and filtrations on Chow groups.* In: *Motives—Seattle 1991* (U. Jannsen, S. Kleiman, J.-P. Serre, eds.), 245–302, Proc. Sympos. Pure Math., 55, Part 1, AMS, Providence 1994.

[49] D. Johnson: *A survey of the Torelli group,* Contemp. Math. 20 (1983), 165–179.

[50] N. Kawazumi, S. Morita: *The primary approximation to the cohomology of the moduli space of curves and cocycles for the stable characteristic classes,* Math. Res. Lett. 3 (1996), 629–641.

[51] S. Keel: *Basepoint freeness for big line bundles in positive characteristic, with applications to* $\mathcal{M}_g$ *and to 3-fold MMP.* Preprint 1997, math.AG/9703003, to appear in Annals of Math.

[52] J. Kollár: *Rational curves on algebraic varieties.* Ergebnisse 32, Springer, Berlin-New York, 1996.

[53] M. Kontsevich: *Intersection theory on the moduli space of curves and the matrix Airy function,* Comm. Math. Phys. 147 (1992), 1–23.

[54] M. Kontsevich: *Enumeration of rational curves via torus actions.* In *The moduli space of curves,* R. Dijkgraaf, C. Faber, G. van der Geer, editors, Birkhäuser, 1995, 335–368.

[55] F.F. Knudsen: *The projectivity of the moduli space of stable curves. II. The stacks* $M_{g,n}$. Math. Scand. 52 (1983), no. 2, 161–199.

[56] J. Li and G. Tian, *Virtual moduli cycles and Gromov-Witten invariants of algebraic varieties,* Jour. AMS 11 (1998), no. 1, 119–174.

[57] E. Looijenga: *Smooth Deligne-Mumford compactifications by means of Prym-level structures,* J. Alg. Geom., 3 (1994), 283–293.

[58] E. Looijenga: *On the tautological ring of* $\mathcal{M}_g$. Invent. Math. 121 (1995), no. 2, 411–419.

[59] G. Mess: *The Torelli groups for genus 2 and 3 surfaces*, Topology 31 (1992), 775–790.

[60] E. Miller: *The homology of the mapping class group*, J. Diff. Geom. 24 (1986), 1–14.

[61] S. Morita: *Characteristic classes of surface bundles*, Invent. Math. 90 (1987), 551–577.

[62] D. Mumford, J. Fogarty, F. Kirwan: *Geometric invariant theory*. Ergebnisse 34, Springer, Berlin-New York, 1994.

[63] D. Mumford: *Curves and their Jacobians*. University of Michigan Press, Ann Arbor 1975.

[64] D. Mumford: *Towards an enumerative geometry of the moduli space of curves*. In: Arithmetic and Geometry (Eds. M. Artin and J. Tate), Vol. II, 271–328, Progress in Math. 36, Birkhäuser, Boston, 1983.

[65] E. Nicorestianu: *On complete subvarieties of* $\mathcal{M}_g$. Thesis, Humboldt Universität, 1998.

[66] F. Oort: *Complete subvarieties of moduli spaces*. In: Abelian varieties (Eds. W. Barth, K. Hulek, H. Lange), 1995, 225–235.

[67] R. Pandharipande: *A geometric construction of Getzler's relation*. Preprint 1997, math.AG/9705016.

[68] M. Pikaart: *An orbifold partition of* $\overline{\mathcal{M}}_g^n$, in *The moduli space of curves*, R. Dijkgraaf, C. Faber, G. van der Geer, editors, Birkhäuser, 1995, 467–482.

[69] M. Pikaart and A.J. de Jong: *Moduli of curves with non-abelian level structure*, in *The moduli space of curves*, R. Dijkgraaf, C. Faber, G. van der Geer, editors, Birkhäuser, 1995, 483–509.

[70] G. Shimura: *Introduction to the arithmetic theory of automorphic functions*. Publications of the Mathematical Society of Japan, No. 11. Iwanami Shoten, Publishers, Tokyo; Princeton University Press, Princeton, N.J., 1971.

[71] J.-L. Verdier: *Sur les intégrales attachées aux formes automorphes (d'après Goro Shimura)*. Séminaire Bourbaki (1960/61), Exp. No. 216.

[72] B. Wajnryb: *A simple presentation for the mapping class group of an orientable surface*, Israel J. Math. 45 (1983), 157–174.

[73] E. Witten: *Two-dimensional gravity and intersection theory on moduli space*. Surveys in differential geometry (Cambridge, MA, 1990), 243–310, Lehigh Univ., Bethlehem, PA, 1991.

[74] C. Zaal: *Explicit complete curves in the moduli space of curves of genus three*. Geom. Dedicata 56 (1995), no. 2, 185–196.

[75] C. Zaal: *A complete surface in* $\mathcal{M}_6$ *in characteristic* $p > 2$. Preprint 1998, available from http://turing.wins.uva.nl/~zaal . To appear in Comp. Math.

[76] H. Zieschang, E. Vogt, H.-D. Coldewey: Surfaces and Planar Discontinuous Groups, Lecture Notes in Math. 835, Springer Verlag (1980).

A Stratification of a Moduli Space of Polarized Abelian Varieties in Positive Characteristic

Frans Oort

Abstract

In this paper we study the moduli space $\mathcal{A}$ of principally polarized abelian varieties of dimension g defined over a field of characteristic p. For moduli spaces one can try to obtain a stratification, by defining a discrete invariant for the objects to be classified, and by taking as strata those loci where the invariant considered is constant. Here we use the observation that finite group schemes annihilated by p geometrically "have no moduli". For every abelian variety X we consider the finite group scheme $X[p]$ (the kernel of multiplication by p). This can be encoded conveniently in the notion of an "elementary sequence". Raynaud proved that the largest stratum, the ordinary locus, is quasi-affine. It is the generalization of that method which makes everything work. In particular we show that every stratum in the EO-stratification is quasi-affine. A careful study of the way strata attach to each other gives a connectedness result. This generalizes a result by Faltings and by Chai which says that $\mathcal{A}$ is irreducible. This is joint work with T. Ekedahl.

1 Introduction

This is a survey of joint work with T. Ekedahl. For details we refer to the paper [6]. We fix some notations. Let $g \in \mathbb{Z}_{>0}$. We consider abelian varieties of dimension g. We fix a prime number p, and a positive integer n not divisible by p. We denote by $\mathcal{A}_{g,1,n}$ the moduli scheme over $\mathrm{Spec}(\mathbb{Z}[1/n])$ of principally polarized abelian varieties of dimension g with a symplectic level-n-structure. We write

$$\mathcal{A} := \mathcal{A}_{g,1,n} \otimes \mathbb{F}_p$$

for the related moduli space of principally polarized abelian varieties defined in characteristic p.

The main result of [6] is the construction of a *stratification* of $\mathcal{A}$ such that the closure of every positive dimensional stratum is connected. This proves *irreducibility* of the moduli space $\mathcal{A}$, a result which was proved by G. Faltings and by C.-L. Chai, see [7], page 364; [1], page 191, Coroll. 4.2.2; [8], page 131, 6.8, Coroll. 1; in their proof irreducibility of $\mathcal{A}_{g,1,n} \otimes \mathbb{C}$ was used; our method lives completely in positive characteristic; we derive irreducibility of all geometric fibres of $\mathcal{A}_{g,1,n} \to \mathrm{Spec}(\mathbb{Z}[1/n])$.

In positive characteristic one can think of the locus of *ordinary* abelian varieties as the interior of $\mathcal{A}$, and one can consider the "boundary" as that locus where either the abelian variety degenerates, or where the p-rank drops. This method has several nice aspects, it seems to offer a strong substitute for complex analytic and topological methods in cases where those are not available. As an application we mention work of C.-L. Chai, where denseness of the Hecke orbit of an ordinary abelian variety is proved, see [2].

The stratification we construct "depends on p" in the sense that there is no stratification of $\mathcal{A}_{g,1,n}$ reducing mod p to this stratification for every p. However, the Chow classes defined by the various strata make perfectly good sense in characteristic zero. T. Ekedahl and G. van der Geer have given another interpretation of the stratification described in this paper, and they compute the Chow classes defined by these strata, see [9] in this volume. We expect several applications from this approach. We expect that certain results in characteristic zero can be proved in this way, using the geometry in characteristic p.

As is usual in working on moduli spaces there is an interplay between local and global methods. We construct our stratification by "point-wise" considerations: the moduli-interpretation of every point will determine on which stratum it is situated. Dimensions are computed in a local fashion, using deformation theory. The global aspect comes from ampleness of the determinant of the dualizing sheaf, well-known in characteristic zero, and proved by Moret-Bailly in all characteristics. Long ago Raynaud observed that a stable ordinary curve over a complete curve in characteristic p is isotrivial, see Szpiro [32], Th. 5, page 62. This was generalized to families of abelian varieties by Moret-Bailly, see [22], XI.5, page 237. The stratification we construct allows us for *principally polarized* abelian varieties to use the same method for cases where the "canonical filtration" jumps. This is the essential tool for us to show that the strata we construct are quasi-affine (i.e. open in an affine scheme). We finish by showing that the closure L of the one-dimensional stratum is in the supersingular locus, and we show that this one-dimensional closed set is connected.

2 Elementary Sequences and Final Filtrations

Definition 2.1. An *elementary sequence* φ

$$\{\varphi(1), \ldots, \varphi(g)\},$$

is a sequence $\varphi \in \mathbb{Z}^g$, we shall write $\varphi(0) = 0$, such that

$$\varphi(i) \leq \varphi(i+1) \leq \varphi(i) + 1 \quad \text{for} \quad 0 \leq i < g.$$

The set of elementary sequences is denoted by Φ. Clearly $\#(\Phi) = 2^g$. We write

$$|\varphi| := \sum_{i=0}^{i=g} \varphi(i),$$

a number called the *dimension* of φ. The sequence $\{0, \cdots, 0\}$ is called the *super-special sequence*. The sequence $\{1, 2, \cdots, g\}$ is called the *ordinary sequence*.

For the ordinary elementary sequence $\varphi = \{1, 2, \cdots, g\}$ we write $f(\varphi) = g$. Otherwise, for an elementary sequence $\varphi \in \Phi$ we write $f = f(\varphi)$ for the (non-negative) integer with the property $\varphi(f) = f = \varphi(f+1)$. We write $a = a(\varphi) := g - \varphi(g)$. The significance of these integers will be described below.

2.2. The kernel of $[p]$. Let (X, λ) be a principally polarized abelian scheme, and let m be a positive integer. Consider the finite group scheme

$$X[m] := \text{Ker}(\times m : X \to X).$$

The principal polarization equips $X[m]$ with a non-degenerate pairing. This can be seen as follows. The exact sequence

$$0 \to N = X[m] \to X_1 \xrightarrow{m} X_2 \to 0$$

(for convenience we have numbered the copies of X in consideration) gives by the duality theorem (see [24], Th. 19.1) exactness of the sequence

$$0 \to N^D = X_2^t[m] \to X_2^t \xrightarrow{m} X_1^t \to 0.$$

The isomorphism $\lambda : X_1 \to X_2^t$ gives an isomorphism

$$N = X_1[m] \xrightarrow{\sim} X_2^t[m] = N^D,$$

and this is the duality (the pairing) on $X[m]$ we are looking for.
We apply this in case (X, λ) is principally polarized and $m = p$ a prime number (which happens to be the characteristic of the base field):

$$(X, \lambda) \quad \mapsto \quad (X[p], <, >).$$

2.3. F and V. (See [30], Exp. VII$_A$ by P. Gabriel, especially VII$_A$.4: "Frobeniuseries".) For any scheme Y over $\text{Spec}(\mathbb{F}_p)$ we have the absolute Frobenius $f : Y \to Y$ (on affine schemes this is exponentiation by p for all elements in the coordinate ring). For a scheme $S \to \text{Spec}(\mathbb{F}_p)$ and a relative scheme $X \to S$ we define $X^{(p/S)} = X^{(p)} \to S$ by the cartesian diagram (fibre product):

$$X^{(p)} \longrightarrow X$$
$$\downarrow \qquad\qquad \downarrow$$
$$S \xrightarrow{\;f\;} S.$$

The absolute Frobenius $f : X \to X$ factors through an S-morphism $F : X \to X^{(p)}$, the "Frobenius morphism",

$$f = (X \xrightarrow{F} X^{(p)} \to X).$$

For a *commutative* group scheme $G \to S$ in characteristic p there is a natural S-homomorphism

$$V : G^{(p)} \to G,$$

the "Verschiebung", and $F \cdot V = p$, and $V \cdot F = p$.

In case $G = X[p]$ is the p-kernel of an abelian scheme X we have

$$\mathrm{Im}(V : G^{(p)} \to G) = \mathrm{Ker}(F : G \to G^{(p)})$$

and

$$\mathrm{Im}(F : G \to G^{(p)}) = \mathrm{Ker}(V : G^{(p)} \to G).$$

A finite group scheme of rank p^{2g} satisfying these properties is called a BT_1 *truncated group scheme*. See [11] for a discussion of such group schemes. In the case where moreover there is an isomorphism

$$\lambda : G \to G^D \quad \text{such that} \quad \lambda^D \cdot \lambda = \mathrm{id}_G ,$$

the pair (G,λ) is called a *polarized BT_1 truncated group scheme*; here G^D stands for the Cartier dual of the commutative finite group scheme G, see [24], I.2. Every such is the truncation of a polarized p-divisible group. We will indicate the isomorphism $\lambda : G \to G^D$ as a perfect pairing $< , >: G \times G \to \mu_p$. We write

$$H := \mathrm{Im}(V : G^{(p)} \to G).$$

In case $(G, < , >)$ is a polarized BT_1 truncated group scheme, $H = \mathrm{Im}(V)$ is maximal isotropic, i.e. $H = H^\perp$.

Definition 2.4. A sequence

$$\{\psi(1), \cdots , \psi(g); \psi(g+1), \cdots , \psi(2g) = g\}$$

is called a *final sequence* if $\varphi := \{\psi(1), \cdots , \psi(g)\}$ is an elementary sequence, and if

$$\psi(i) + g - i = \psi(2g - i), \quad 0 \le i \le g.$$

Note that any elementary sequence determines uniquely a final sequence and conversely.

Definition 2.5. Let $(G, <, >)$ be a polarized BT_1 truncated group scheme of rank p^{2g}. A *final filtration* is a filtration

$$0 = G_0 \subset \cdots \subset G_i \subset \cdots \subset G_g = H \subset \cdots \subset G_{2g} = G$$

such that there exists a final sequence ψ, and such that the following properties are satisfied:

$$\mathrm{rk}(G_j) = p^j, \quad G_j^{\perp} = G_{2g-j}, \quad \forall j,$$

(i.e. this is a "symplectic flag") and

$$\mathrm{Im}(V : G_j^{(p)} \to G) = G_{\psi(j)}, \quad \forall j;$$

in particular $H = G_g = V(G_{2g}^{(p)})$.

2.6. Starting from (X,λ) we try to find a final filtration. It turns out we can find canonically something weaker; it will be called a canonical filtration. Over an algebraically closed field it can be refined to a final filtration. That filtration is not unique in general, but for a given (X,λ) the final sequence (and hence the elementary sequence) turns out to be unique.

First we explain the construction of the *canonical type*. Suppose given a principally polarized abelian variety (X,λ) over a field K (of positive characteristic). We start with

$$0 \subset H = \mathrm{Im}(V) \subset G := X[p].$$

We take the images under V of all subgroups appearing; this only produces possibly new subgroups between $0 \subset H$; of the new sequence of subgroups we add all duals; this produces possibly new subgroups between $H \subset G$. With the new set of subgroups we repeat this process. After a finite number of such steps we arrive at a sequence of subgroups of G

$$0 \subset C_1 \subset \cdots \subset C_t = H \subset \cdots \subset C_{2t} = G$$

which is symplectic and stable under V, such that each of them is obtained by applying a finite number of times V or $\perp$ starting from G; this is unique once $(G, <, >)$ is given, it is called the *canonical sequence* of (X,λ).

An important aspect of a canonical sequence is the following: for every index i, the map V is zero on $C_{i+1}^{(p)}/C_i^{(p)}$ if and only if it is injective on $C_{2t-i}^{(p)}/C_{2t-i-1}^{(p)}$.

Theorem 2.7. *For a principally polarized abelian variety (X,λ) over a field K its canonical type can be refined over an algebraic extension of K to a final type; the number of final filtrations is finite, the final sequence obtained is unique, once (X,λ) is given.*

Notation 2.8. The elementary sequence φ thus obtained will be denoted by $\varphi = \mathrm{ES}(X,\lambda) = \mathrm{ES}(X[p], <, >)$.

Remark 2.9. We shall see that every $\varphi \in \Phi$ can be obtained in this way.

2.10. The p-rank. For an abelian variety X in characteristic p we define its p-rank $f = f(X)$ by:
$$\text{Hom}(\mu_p, X \otimes k) \cong (\mathbb{Z}/p)^f;$$
here k is an algebraically closed field. For any principal polarization λ on X and $\varphi = \text{ES}(X, \lambda)$ we have $f(X) = f(\varphi)$. The same definition can be given for a semi-abelian variety. Note that for an abelian variety

$$X[p](k) = \text{Hom}(\mathbb{Z}/p, X \otimes k) \cong (\mathbb{Z}/p)^f \cong \text{Hom}(\mu_p, X \otimes k).$$

An abelian variety (or a semi-abelian variety) is called *ordinary* if $f(X) = \dim(X)$.

Note the following fact: if N is a group scheme of rank p over an algebraically closed field of characteristic p, then one of the three possibilities holds: either $N \cong \mu_{p,k}$, or $N \cong \alpha_{p,k}$, or $N \cong (\mathbb{Z}/p)_k$, where over $\mathbb{F}_p$ we define: $\alpha_p := \text{Ker}(F : \mathbb{G}_a \to \mathbb{G}_a)$.

2.11. The a-number. For an abelian variety X over a perfect field K of characteristic p we define
$$a(X) = \dim_K(\text{Hom}(\alpha_p, X)).$$
Note that $a(X) = a(\text{ES}(X, \lambda))$.

Theorem/Notation 2.12. *Let $(\mathcal{X}, \lambda) \to S$ be a principally polarized abelian scheme over a scheme S in characteristic p. For a given $\varphi \in \Phi$ consider the set of all $(X, \lambda) \in S$, defined over some field K, depending on (X, λ), such that $\text{ES}(X, \lambda) = \varphi$. This set, denoted by $\mathcal{C}_\varphi(S)$, is locally closed in S. We write $\mathcal{C}_\varphi(\mathcal{A}) =: S_\varphi$.*

The proof of this theorem uses group schemes over an arbitrary base, the result follows because for a given homomorphism of finite flat group schemes the kernel has a given rank over a locally closed subscheme.

2.13. Remark: EO-strata. Note that we define $S_\varphi \subset \mathcal{A} = \mathcal{A}_{g,1,n} \otimes \mathbb{F}_p$ as the locally closed sets (i.e. the *open* strata) where the elementary sequence equals φ. In some other stratifications (e.g. the NP-stratification, see section 8) one considers the closed strata.

3 Explicit Description of Some of the Strata

3.1. Let $\varphi = \{0, \cdots, 0\}$, i.e. what we call the *superspecial* elementary sequence. We write
$$\Sigma = S_{\{0,\cdots,0\}}.$$

3.2. Fact: For an algebraically closed field k of characteristic p we have: $[(X,\lambda)] \in \Sigma(k)$ if and only if $X \cong E^g$, where E is a supersingular elliptic curve (here we use results by Eichler, Deligne and Oort). The set $\Sigma = S_{\{0,\cdots,0\}}$ has dimension zero, and the number of points in Σ (depends on p and depends on g and) can be given by a class number.

The closure of the stratum $S_{\{0,\cdots,0,1\}}$ is called L:

$$L := (S_{\{0,\cdots,0,1\}})^c.$$

Theorem 3.3. *Let $g > 1$. The closed set $L \subset \mathcal{A}$ is one-dimensional,*

$$L = S_{\{0,\cdots,0,1\}} \cup \Sigma,$$

and the set L is **connected.**

The proof of this theorem involves a careful study of all possible (X,λ) with $\varphi(g) = 1$. Note that in this case $a(X) = g - 1$. It turns out that the types $\varphi = \{0,\cdots,0,1,\cdots,1\}$ with $|\varphi| \leq g/2$ correspond with supersingular abelian varieties with $a(X) = g - 1$. These were studied in [18]. For each of the given possibilities the related φ is computed, only one type gives $\varphi = \{0,\cdots,0,1\}$. For this type we show that for any point in Σ at least one component of L passes through that point, and that every component of L has dimension one (here we use a construction by Moret-Bailly, see [21]). A careful study of perfect hermitian lattices over a quaternion algebra, using strong approximation, see [16], [29], finishes the argument.

4   Standard Types

We consider pairs $(G, < , >)$, where G is a finite group scheme over a perfect field of positive characteristic p, and where $< , >$ is a non-degenerate pairing on its Dieudonné module. We ask ourselves, what kind of group schemes with a bilinear form $(G, < , >)$ can appear as $(X,\lambda)[p]$? Note the following confusing situation: first order deformations of (X_0,λ_0) and of $(X_0,\lambda_0)[p]$ correspond one-to-one (by rigidifying the central fibres, this is a theorem proved by Grothendieck, see Illusie, [11], Th. 4.4). Hence one might expect that there are "many" polarized BT_1 truncated group schemes; there exist rings over which this is true. However, over an algebraically closed ground field the number of such isomorphism classes of a given rank turns out to be finite.

4.1. Construction of standard types. Suppose given a prime number p, an integer $g \in \mathbb{Z}_{>0}$, and an elementary sequence $\varphi \in \Phi$. We construct a polarized BT_1 truncated group scheme $(G_\varphi, < , >)$ (which we we call the standard type given by φ) defined over $\mathbb{F}_p$.

In fact this is given by a Dieudonné module of dimension $2g$ over $\mathbb{F}_p$, with base vectors which give a symplectic base, and such that for every base vector the image under Frobenius or Verschiebung is either 0 or one of the base vectors. It is easy to see how to construct this module from φ.

Here is an easy example, which might explain the basic idea: suppose $\varphi = \{0,1,1\}$, then $\psi = \{0,1,1;2,2,3\}$, we choose a symplectic base $\{Y_3, X_1, Y_2, X_2, Y_1, X_3\}$ (the vectors X_i at the places where the ψ-sequence jumps), and we use ψ to define: $V(Y_i) = 0$ for all i, and $V(X_3) = Y_2$, $V(X_2) = X_1$, $V(X_1) = Y_3$. This fixes F, in fact $F(X_3) = -Y_1$, $F(Y_1) = Y_2$, $F(X_2) = -Y_3$, etc.

Theorem 4.2. *Suppose k is an algebraically closed field. For every polarized BT_1 truncated group scheme $(G, <, >)$ defined over k with $\mathrm{ES}(G, <, >) = \varphi$ there exists an isomorphism*

$$(G, <, >) \quad \cong \quad (G_\varphi, <, >) \otimes_{\mathbb{F}_p} k.$$

[Here $<, >$ is a skew pairing on the Dieudonné module of G.]

The proof is a matter of linear algebra. The Dieudonné module M of a polarized BT_1 truncated group scheme $(G_\varphi, <, >)$ is a vector space with actions of Frobenius and Verschiebung, with a skew pairing on M (having certain properties). The elementary sequence tells which subgroups in a final filtration are mapped to which subgroups. Over an algebraically closed field a choice for base vectors M can be done in such a way that these base vectors form a symplectic base, and such that for each base vector the image under Frobenius or Verschiebung is either 0 or one of the base vectors, where the combinatorics of this "game" is given by φ. This shows it is isomorphic over k with a standard type $(G_\varphi, <, >)$ which is defined over $\mathbb{F}_p$, and it proves the theorem.

The theorem says that the strata S_φ are given by the geometric isomorphism classes of $(X, \lambda)[p]$.

5 Moving in a Stratum

In this section we suppose that $n \geq 3$, thus ensuring that local deformation theory injects into the moduli space $\mathcal{A}$.

Notation 5.1. For $\varphi_1, \varphi_2 \in \Phi$ we write $\varphi_1 \prec \varphi_2$ if for every $1 \leq i \leq g$ we have $\varphi_1(i) \leq \varphi_2(i)$.

Deformation theory of p-divisible groups can be described (say, in equal characteristic) by the method of *displays*. This was invented by Mumford. It was described and used in [23]. A more complete description can be found in [33]. Using moreover the theorem of Serre and Tate, see [15], Section 1, which says that the infinitesimal deformation theory of an abelian variety in residue characteristic and the infinitesimal theory of its p-divisible group are equivalent, we obtain in this way a description of the deformation theory of (polarized) abelian varieties. The advantage of the description using displays is the fact that deformations are not only described by infinitesimal methods, but for a given special fibre (X_0, λ_0) the whole infinitesimal deformation space is written out in coordinates in a complete local ring. Using these methods one can show:

Theorem 5.2. *a) Given $(X_0, \lambda_0) \in \mathcal{A}$. Let $\varphi' = \mathrm{ES}(X_0, \lambda_0)$. We write $d' := |\varphi'|$ for the dimension of φ'. There exists a formally smooth, closed subscheme of dimension d' in the local deformation space of (X_0, λ_0) on which the elementary type of every fibre equals φ'.*

b) Suppose given the data as in (a). Suppose there exists an elementary sequence φ and an index i with $1 \leq i \leq g$ such that $\varphi'(j) = \varphi(j)$ for $j \neq i$, and $\varphi'(i) + 1 = \varphi(i)$ (i.e. $\varphi' \prec \varphi$ and $|\varphi'| + 1 = |\varphi|$). Then there exists a complete local discrete valuation ring R, a family $(\mathcal{X}, \lambda) \to \mathrm{Spec}(R)$ with closed fibre isomorphic to (X_0, λ_0), and with generic fibre having elementary type φ.

This theorem is proved by writing out carefully the deformation theory. The first part is standard. In the second part one has to make a careful choice for a final filtration on $\mathcal{X}[p]$.

Corollary 5.3. *For every $\varphi \in \Phi$ there exists an $(X, \lambda) \in \mathcal{A}$ with $\mathrm{ES}(X, \lambda) = \varphi$.*

Corollary 5.4. *Suppose $n \geq 3$, and prime to p. For every $\varphi \in \Phi$, the stratum $S_\varphi \subset \mathcal{A} = \mathcal{A}_{g,1,n} \otimes \mathbb{F}_p$ is non-singular, and*

$$\dim(S_\varphi) \quad = \quad |\varphi|$$

(finally!, an explanation why we call this "the dimension of φ").

By the first part of the theorem we see that for every φ each component of S_φ has dimension at least $|\varphi|$. By the second we see that any point in a component V' of $S_{\varphi'}$ is in the boundary of a component V of S_φ (notation as in the second part of the theorem), hence $\dim(V') < \dim(V)$ in this case. Note that $\dim(\mathcal{A}) = g(g+1)/2$, and note that a maximal chain of comparable mutually different elementary sequences

$$\varphi = \varphi_0 \prec \varphi_1 \prec \cdots \prec \varphi_t = \{1, 2, \cdots, g\}$$

has exactly length $t = (g(g+1)/2) - |\varphi|$. This proves the dimension statement, and by the first part of the theorem it proves the fact that each S_φ is non-singular.

Theorem 5.5. *Let $\varphi, \varphi' \in \Phi$, and suppose that*

$$S_{\varphi'} \bigcap (S_\varphi)^c \neq \emptyset$$

i.e. the stratum $S_{\varphi'}$ meets the Zariski closure of S_φ. Then for every irreducible component $W' \subset S_{\varphi'}$ there exists an irreducible component $W \subset S_\varphi$ such that $W' \subset W^c$.

A sketch of the proof. Let

$$x_0 = (X_0, \lambda_0) \in S_{\varphi'} \bigcap (S_{\varphi})^c$$

be defined over an algebraically closed field k. Then there exists a complete local ring R and a family $(\mathcal{X}, \lambda)$ over $\mathrm{Spec}(R)$ with special fibre x_0 and generic fibre having elementary sequence φ. We derive a family $(\mathcal{X}, \lambda)[p] = (\mathcal{G}, <, >) \to \mathrm{Spec}(R)$. Let $y_o \in S_{\varphi'}$. There exists a covering T_0 of a Zariski neighborhood of $y_0 \in S_{\varphi'}$ over which the polarized BT_1 truncated group scheme is constant (use the theorem describing standard types). The family $(\mathcal{G}, <, >) \to \mathrm{Spec}(R)$ extends to a family $(\mathcal{G}, <, >) \to T$ where $T_0 \subset T$ is a nilimmersion. Using a theorem of Grothendieck, see Illusie, [11], Th. 4.4.1, we conclude that the p-divisible group along T_0 can be extended to T. Using a theorem of Serre and Tate ([15], Section 1) we conclude that $T_0 \to S_{\varphi'}$ extends to a morphism $T \to (S_{\varphi})^c$. This proves the theorem.

Corollary 5.6. *If R is a complete local discrete valuation ring, and if $(\mathcal{X}, \lambda) \to \mathrm{Spec}(R)$ is a principally polarized abelian scheme where the general fibre has elementary sequence φ and the closed fibre has elementary sequence φ', then*

$$either \quad \varphi' = \varphi \quad or \quad |\varphi'| < |\varphi|.$$

By the theorem we conclude: for an irreducible component $V' \subset S_{\varphi'}$, either $V' \subset S_{\varphi}$, and $\varphi' = \varphi$, or $\dim(V') < \dim(S_{\varphi})$. This proves the corollary.

Remark 5.7. We see that we truly have a stratification, i.e. the boundary of one stratum is the union of strata.

Remark 5.8. Suppose we are in the situation $(\mathcal{X}, \lambda) \to \mathrm{Spec}(R)$ as described in the corollary. In general this does not imply that $\varphi' \prec \varphi$. For example it can be shown that for $g = 7$ the elementary sequence $\varphi' = \{0,0,1,1,2,2,2\}$ shows up in the boundary of S_{φ} with $\varphi = \{0,0,0,1,2,3,3\}$. Indeed $\dim(V') < \dim(S_{\varphi})$ but not $\varphi' \prec \varphi$. Many more examples can be given.

In general, given $\varphi \in \Phi$ we have no reasonable guess which $\varphi' \in \Phi$ appear exactly in the boundary of S_{φ}.

See Van der Geer's paper [9] in this volume for further explanation of these cases, using Weyl groups.

6 The Raynaud Trick

Up to now we have considered local questions. Now it is time to involve global arguments. In order to illustrate what we are going to do, we start with two examples. The first one is the form in which Raynaud, Szpiro and Moret-Bailly formulated this idea. Recall that a semi-abelian variety X is called *ordinary* if $f(X) = \dim(X)$. This is the case if and only if the abelian part is ordinary.

6.1. The Raynaud trick for ordinary abelian varieties. (Raynaud, Szpiro, Moret-Bailly) *Let S be an irreducible scheme in characteristic p, and let $(X,\lambda) \to S$ be a polarized abelian scheme. Suppose that all fibres of X/S are ordinary. Then the Satake map*

$$g : S \to \mathcal{A}_{g,d} \otimes \mathbb{F}_p \quad \text{defined by} \quad (\mathcal{X},\lambda) \to S$$

has an image $g(S) \subset \mathcal{A}$ which is quasi-affine.

For a proof see [22], XI.5, page 237, Th. 5.2. Let us sketch the most important detail of the proof. By taking a level structure, and taking the image, it suffices to show the case $S \subset \mathcal{A}_{g,d,n}$. Note that the kernel $\mu := \mathcal{X}[F] \to S$ has the same relative tangent space as $\mathcal{X} \to S$. Note that every geometric fibre of $\mu \to S$ is isomorphic with $(\mu_p)^g$. Hence there exists a finite cover $T \to S$ which trivializes it:

$$(\mu \to S) \times_S T \cong (\mu_{p,T})^g.$$

This shows that $\mathrm{Lie}(\mathcal{X}_T/T)$ is a trivial vector bundle on T, hence $\mathrm{Det}(\mathrm{Lie}(\mathcal{X}/S))$ is a torsion line bundle on S. Consider the universal abelian scheme $\mathcal{Y} \to \mathcal{A}_{g,d,n} \otimes \mathbb{F}_p =: \mathcal{B}$. We know by Moret-Bailly that

$$\omega = \mathrm{Det}((\mathrm{Lie}(\mathcal{Y}/\mathcal{B}))^{dual}$$

is ample. Hence the image of S is quasi-affine.

This shows that every irreducible component of the ordinary stratum is quasi-affine; hence this is true for every component of S_φ with $\varphi = \{1,2,\cdots,g\}$. Note that we did not use any information about the degree of the polarization. As another example we treat the next stratum, and here we use that we work with *principal* polarizations.

6.2. The case of p-rank $= g-1$. *Let S be an irreducible scheme in characteristic p, and let $(\mathcal{X},\lambda) \to S$ be a principally polarized abelian scheme. Suppose that all fibres of $\mathcal{X}/S$ have p-rank equal to $g-1$, i.e. every fibre has elementary sequence equal to $\varphi = \{1,2,\cdots,g-1,g-1\}$. Then the Satake map*

$$g : S \to \mathcal{A}_{g,1} \quad \text{defined by} \quad (\mathcal{X},\lambda) \to S$$

has an image $g(S) \subset \mathcal{A}$ which is quasi-affine.

This is the central idea which makes the stratification we are studying useful. Let us sketch the proof. Taking a level structure it suffices to show the case $S \subset \mathcal{A} = \mathcal{A}_{g,1,n} \otimes \mathbb{F}_p$. Consider $\mathcal{X}[p] \to S$. Because the p-rank is constant, equal to $g-1$ on S, there exists an exact sequence of flat finite group schemes

$$0 \to \mu \to \mathcal{X}[F] \to \mathcal{N}_1 \to 0,$$

where on each fibre $\mathcal{X}[F]_s$ we have that μ is the local-étale and $\mathcal{N}_1$ is the local-local part of $\mathcal{X}[F]$. For μ arguments as above can be applied, showing that $\bigwedge^{g-1}(\mathrm{Lie}(\mu/S))$ is a torsion-line bundle on S. However, the group scheme $\mathcal{N}_1$ has fibres isomorphic to α_p. Note that α_p does not have a "canonical generator", how do we see that the line bundle $\mathrm{Lie}(\mathcal{N}_1/S)$ is torsion? Over a finite cover $T \to S$ we have

$$\mathcal{X}[p] \times_S T \cong \mu_T \times \mathcal{N}_T \times (\mu_T)^D,$$

where μ, respectively $\mathcal{N}$, respectively μ^D are the local-étale, resp. local-local, resp. the étale-local part of $\mathcal{X}[p]$. Note that

$$\mathrm{Lie}(\mathcal{N}_T) \subset \mathrm{Lie}(\mathcal{X}_T) \quad \text{and} \quad \mathrm{Lie}(\mathcal{X}_T) = \mathrm{Lie}(\mu_T) \oplus \mathrm{Lie}(\mathcal{N}_T).$$

We have an exact sequence of finite flat group schemes over S:

$$0 \to \mathcal{N}_1 \to \mathcal{N} \to \mathcal{N}_2 \to 0.$$

By duality we have $\mathcal{N}^D = (\mathcal{X}^t[p])_{loc,loc}$, and hence

$$\lambda_{|\mathcal{N}_1} : \mathcal{N}_1 \xrightarrow{\sim} \mathcal{N}_2^D$$

is an isomorphism (here we use that we work with a principal polarization). The morphism $V : \mathcal{N}^{(p)} \to \mathcal{N}$ defines an isomorphism $V : \mathcal{N}_2^{(p)} \to \mathcal{N}_1$. We see that λ and V together define:

$$(\mathrm{Lie}(\mathcal{N}_1/S))^{\otimes -p} \xrightarrow{\sim} (\mathrm{Lie}(\mathcal{N}_2^D/S))^{\otimes -p} = (\mathrm{Lie}(\mathcal{N}_2/S))^{\otimes p} \xrightarrow{\sim} \mathrm{Lie}(\mathcal{N}_1/S);$$

hence

$$(\mathrm{Lie}(\mathcal{N}_1/S))^{\otimes(-1-p)} = 1.$$

As $\mathrm{Lie}(\mathcal{X}_T) = \mathrm{Lie}(\mu_T) \oplus \mathrm{Lie}(\mathcal{N}_T)$ this shows, together with the previous argument applied to μ that $\mathrm{Det}(\mathrm{Lie}_{\mathcal{X}/S})$ is torsion, and we finish as before. This shows that the stratum $S_{\{1,\cdots,g-1,g-1\}}$ is quasi-affine.

6.3. The stratification $\{S_\varphi\}_{\varphi \in \Phi}$ is made in such a way that the previous arguments apply to every stratum. The images produced by a finite sequence of maps V and duality $\perp$ give flat group schemes of finite (constant) rank on one stratum. Hence the previous arguments can be used. Moreover this can be done on the Satake compactification $\mathcal{A}^*$ of $\mathcal{A} = \mathcal{A}_{g,1,n}$ (the minimal compactification described in [8]).

We define the elementary sequence of a semi-abelian variety in the following way: if (X',λ') is a semi-abelian variety with a principal cubic structure (over an algebraically closed field), and $(\mathbb{G}_m)^s = L \subset X$ is the maximal connected linear subgroup, then λ' induces a principal polarization λ on $X := X/L$; we write $\varphi = \mathrm{ES}(X,\lambda)$, and we define

$$\varphi(X',\lambda') = \varphi' = \{1, 2, \ldots, s, \ldots, \varphi'(i+s) = \varphi(i) + s, \ldots\}.$$

By $\mathcal{A}^*$ we denote the Satake compactification of $\mathcal{A}$, called the minimal compactification in [8].

Theorem 6.4. *Let S be an irreducible scheme in characteristic p. Let $(\mathcal{X},\lambda) \to S$ be a semi-abelian scheme over S with a principal cubic structure with generic fibre an abelian variety. Suppose that the canonical stratification in this family is constant, i.e. there exists $\varphi \in \Phi$ such that $C_\varphi(S) = S$. Then the Satake map*

$$g : S \to \mathcal{A}^* \quad \text{defined by} \quad (\mathcal{X},\lambda) \to S$$

has an image $g(S) \subset \mathcal{A}^*$ which is quasi-affine.

6.5. For $\varphi \in \Phi$ we define the stratum $T_\varphi \subset \mathcal{A}^*$ to be the set where the related semi-abelian variety has elementary sequence φ. One shows that this is a locally closed set in $\mathcal{A}^*$. Clearly $S_\varphi \subset T_\varphi$. By the previous result one shows that for every φ the set T_φ is quasi-affine. One shows that these strata are "transversal to the boundary" (in a well-chosen toroidal compactification), in particular, every boundary point with $\mathrm{ES}(X_0,\lambda_0) = \varphi$ is in the closure $(S_\varphi)^*$.

Notation 6.6. For a given integer d with $0 \le d \le g(g+1)/2$ we write

$$\bigcup_{|\varphi|=d} T_\varphi = \mathbb{T}_d \subset \mathcal{A}^*.$$

7   Results

Proposition 7.1. *For every d, the closed subset*

$$(\mathbb{T}_{d-1})^* \subset (\mathbb{T}_d)^*$$

is an ample divisor.

[The star denotes: closure in the Satake compactification.]

Theorem 7.2. *For every $\varphi \in \Phi$ the stratum*

$$S_\varphi \subset \mathcal{A} = \mathcal{A}_{g,1,n} \otimes \mathbb{F}_p$$

is quasi-affine, and

$$\dim(S_\varphi) = |\varphi|.$$

The disjoint union

$$\sqcup_{\varphi \in \Phi} S_\varphi = \mathcal{A}$$

is a stratification, i.e. for every φ the boundary of S_φ is a union of strata. For $n \ge 3$ every stratum S_φ is regular.

Note that we write $L = (S_{\{0,\cdots,0,1\}})^c$ for the closure of the *unique* one-dimensional stratum.

Corollary 7.3. *For every $\varphi \in \Phi$ with $\varphi \neq \{0, \cdots, 0\}$*

$$L \subset (S_\varphi)^c \quad and \quad (S_\varphi)^c \quad is \ connected.$$

Assume $g > 1$. In fact, as we have seen, the closure of every irreducible component of every S_φ contains an irreducible component of a lower stratum. Hence for every irreducible component V of $\varphi \in \Phi$ we have, using the theorem

$$V^c \cap \Sigma \neq \emptyset, \quad and \quad L \subset (S_\varphi)^c.$$

As L is connected this proves the claim.

Corollary 7.4. *(G. Faltings, C.-L. Chai) For every $g \in \mathbb{Z}_{>0}$ and every $n \in \mathbb{Z}_{>0}$*

$$\mathcal{A}_{g,1,n} \otimes \overline{\mathbb{F}_p} \quad is \ irreducible.$$

In fact, for $n \geq 3$ this scheme is regular, and it is connected by the previous corollary; hence it is irreducible. Then it follows for all $n > 0$.

Remark 7.5. This result was proved by Faltings and by Chai using information obtained from methods in characteristic zero. Note that our proof uses only methods in positive characteristic; moreover irreducibility of $\mathcal{A}_{g,1} \otimes \mathbb{C}$ follows from this.

Notation 7.6. We write $V_{g-1} \subset \mathcal{A}$ for the locus where the abelian variety is not ordinary. In fact,

$$V_{g-1} = (S_{\{1,\cdots,g-1,g-1\}})^c = (\mathbb{S}_{(g(g+1)/2)-1})^c.$$

Corollary 7.7. *For $g > 1$*

$$V_{g-1} \quad is \ irreducible.$$

From the previous arguments it follows that $(V_{g-1})^c \supset L \supset \Sigma$. An easy computation in local coordinates around a superspecial point shows that locally at such a point the germ of $(V_{g-1})^c$ is given by $\mathrm{Det}(T_{i,j}) = 0$, where $T_{i,j} = T_{j,i}$, $1 \leq i \leq j \leq g$ are local parameters. For $g > 1$ this equation defines an irreducible germ. Hence for every point $x \in \Sigma$ there is precisely one irreducible component of $(V_{g-1})^c$ containing x. By the connectedness of L this shows irreducibility of $(V_{g-1})^c$.

Remark 7.8. A generalization of the corollary can be found in [9], Theorem 9.9.

8 Some Questions

8.1. Irreducibility of strata. In general $\Sigma = \mathbb{S}_0 = S_{\{0,\cdots,0\}}$ and $L = (\mathbb{S}_1)^c = (S_{\{0,\cdots,0,1\}})^c$ have many components, e.g. see [12], Th. 2.10 and computations by Eichler, by Hashimoto and Ibukiyama.

Take $r = [g/2]$; consider the stratum $S_{\{0,\cdots,0,1,\cdots,1\}}$, where the number 1 is appearing exactly r times. This has many components in general, see [18], Prop. 9.11.

As an example, for $g = 3$ and $\varphi = \{0,1,1\}$ we know that the related $S_\varphi \subset \mathcal{A}_3$ is irreducible, while in general $S_{\{0,0,1\}}$ is reducible.

Conjecture 8.2. *We expect that S_φ is irreducible in those cases where a generic point does not correspond with a supersingular abelian variety.*

In [25], we study a stratification of $\mathcal{A}$ by Newton polygons. These strata are closed subsets of $\mathcal{A}_g$. [We have adopted the convention that the NP-strata are closed: take all polarized abelian varieties with NP lying on or above the given one; the EO-strata defined above are open: take all principally polarized abelian varieties where the kernel has exactly this elementary sequence.] We know that (for p and g "large") the supersingular stratum has many components, see [18], 4.9.

Conjecture 8.3. *We expect that for every Newton polygon which is not super-singular (i.e. which has at least one slope not equal to 1/2), the related Newton polygon stratum is irreducible.*

Several special cases of this have been proved.

8.4. Hecke orbits. Take some $\varphi \in \Phi$, take $x \in S_\varphi$ and consider the separable Hecke orbit $\mathcal{H}^{(\ell)}(x)$ of x as in [2]. Is

$$\mathcal{H}^{(\ell)}(x) \subset S_\varphi$$

a dense subset? In general the answer is "no", here are two types of examples: if all points of S_φ correspond with a supersingular abelian variety, then this Hecke orbit is finite, hence for every $\varphi \in \Phi$ with $\varphi(g) = 1$, and $|\varphi| \leq [g/2]$ the answer is negative. If several isogeny types are appearing on S_φ the answer is negative in general, e.g. take $0 \leq f \leq g-3$ and φ with $\varphi(g) = g-1$ and $\varphi(f) = f = \varphi(f+1)$.

Conjecture 8.5. *Let $[(X,\lambda)] = x \in \mathcal{A}_g \otimes \mathbb{F}_p$, and let $\beta = \mathcal{N}(X)$ be the corresponding Newton Polygon. Consider the Hecke orbit $\mathcal{H}^{(\ell,p)}(x)$, which is defined by allowing isogenies of ℓ-power order for one fixed prime number $\ell \neq p$, and allowing isogenies which are successive α_p-quotients. We expect that this Hecke orbit is dense in the Newton Polygon stratum $W_\beta \subset \mathcal{A}_g \otimes \mathbb{F}_p$.*

See [26], 16A. Special cases have been proved.

8.6. Newton polygons. The space $\mathcal{A} = \mathcal{A}_{g,1,n} \otimes \mathbb{F}_p$ can be stratified by Newton polygons, see [25]. One can intersect that stratification with the one described in this paper. Some of our strata are contained completely in one NP-stratum, e.g. L is contained in the supersingular stratum. Some EO-strata intersect several of the NP-strata. For example, for $\varphi = \{0,1,2,\cdots,g-1\}$ we see that S_φ meets all NP-strata with $f = 0$, hence for $g > 2$ more than one. One can try to obtain full information about these intersections. We have several examples of an EO-stratum completely contained in one NP-stratum. We have examples of an EO-stratum meeting many NP-strata. We do not have a complete list which determines for which elementary sequence φ and which Newton Polygon β we have $S_\varphi \cap W_\beta^o \neq \emptyset$.

8.7. The Torelli locus. Let $\mathcal{M}_g$ be the moduli space of curves of genus g, and $j : \mathcal{M}_g \to \mathcal{A}_{g,1}$ the Torelli mapping; its image $j(\mathcal{M}_g) = \mathcal{T}_g^o$ is called the (open) Torelli locus. It seems quite hard to describe the intersections of the Torelli locus with strata in the sense described above; the same for intersections of $\mathcal{T}_g^o$ with NP-strata; examples show that these intersections are not transversal in general; it is not clear that dimensions of strata only depend on numerical data describing the strata, it might be that for different values of p we obtain different dimensions. These seem difficult questions, which should be studied more closely.

As an example: in [10] we see that for $p = 2$ and arbitrary g there are supersingular curves. However the longest chain of comparable Newton Polygons, which has length $(g(g+1)/2) - [g^2/4]$, is longer than $\dim(\mathcal{M}_g \otimes K) = 3g - 3$ for $g \geq 9$. We do not know the answer to the following question:

Question 8.8. Consider some value $g \in \mathbb{Z}_{>0}$. Does there exist a Newton Polygon not appearing as the NP of the Jacobian of an algebraic curve of genus g?

We expect that for large g there do exist Newton Polygons not showing up on $\mathcal{M}_g \otimes \mathbb{F}_p$.

Bibliography

[1] C.-L. Chai: *Compactification of Siegel moduli schemes.* London Math. Soc. Lect. Notes Ser. 107, Cambridge Univ. Press, 1985.

[2] C.-L. Chai: *Every ordinary symplectic isogeny class in positive characteristic is dense in the moduli.* Invent. Math. **121** (1995), 439–479.

[3] M. Demazure: *Lectures on p-divisible groups.* Lect. Notes Math. 302, Springer Verlag, 1972.

[4] M. Eichler: *Über die Idealklassenzahl total definiter Quaternionenalgebren.* Math. Z. **43** (1938), 102–109.

[5] M. Eichler: *Zur Zahlentheorie der Quaternionen-Algebren.* J. Reine Angew. Math. **195** (1955), 127–151.

[6] T. Ekedahl, F. Oort: *A stratification of a moduli space of abelian varieties.* In preparation.

[7] G. Faltings: *Arithmetische Kompaktifizierung des Modulraums der abelschen Varietäten.* Arbeitstagung Bonn 1984 (Ed. F. Hirzebruch et al.), Lect. Notes Math. 1111, Springer Verlag, 1985.

[8] G.Faltings, C.-L. Chai: *Degeneration of abelian varieties.* Ergebn. Math. 3.Folge, Bd. 22, Springer Verlag, 1990.

[9] G. van der Geer: *Cycles on the moduli space of abelian varieties.* This volume, 65–89.

[10] G. van der Geer, M. van der Vlugt: *On the existence of supersingular curves of given genus,* J. Reine Angew. Math. **45** (1995), 53–61.

[11] L. Illusie: *Déformations de groupes de Barsotti-Tate.* Exp.VI in: Séminaire sur les pinceaux arithmétiques: la conjecture de Mordell (L. Szpiro), Astérisque 127, Soc. Math. France 1985.

[12] T. Ibukiyama, T. Katsura, F. Oort: *Supersingular curves of genus two and class numbers.* Compos. Math. **57** (1986), 127–152.

[13] J. de Jong, F. Oort: *Purity of the stratification by Newton polygons.* Utrecht Preprint 1081, December 1998; 35 pp.

[14] T. Katsura, F. Oort: *Families of supersingular abelian surfaces.* Compos. Math. **62** (1987), 107–167.

[15] N. M. Katz: *Serre-Tate local moduli.* In: Surfaces algébriques (Sém. de géom. algébr. d'Orsay 1976–78). Lect. Notes Math. 868, Springer Verlag 1981; Exp. V-bis, pp. 138–202.

[16] M. Kneser: *Strong approximation.* Proc. Sympos. Pure Math., Vol. 9: Algebraic groups and discontinuous subgroups. A.M.S. 1966.

[17] N. Koblitz: *p-adic variation of the zeta-function over families of varieties defined over finite fields.* Compos. Math. **31** (1975), 119–218.

[18] K.-Z. Li, F. Oort: *Moduli of supersingular abelian varieties.* Lecture Notes in Mathematics 1680, Springer Verlag, Berlin-New York, 1998.

[19] J. Lubin: *Canonical subgroups of formal groups.* Transact. AMS **251** (1979), 103–127.

[20] W. Messing: *The crystals associated to Barsotti-Tate groups: with applications to abelian schemes.* Lect. Notes Math. 264, Springer Verlag 1972.

[21] L. Moret-Bailly: *Familles de courbes et de variétés abéliennes sur $\mathbb{P}^1$. Familles de courbes et de variétés abéliennes.* Sém. sur les pinceaux de courbes de genre au moins deux. Ed. L. Szpiro. Astérisque 86, Soc. Math. France, 1981; Exp. 7, 8; pp. 109–140.

[22] L. Moret-Bailly: *Pinceaux de variétés abéliennnes.* Astérisque 129, Soc. Math. France, 1985.

[23] P. Norman, F. Oort: *Moduli of abelian varieties.* Ann. Math. **112** (1980), 413–439.

[24] F. Oort: *Commutative group schemes.* Lect. Notes Math. 15, Springer Verlag 1966.

[25] F. Oort: *Moduli of abelian varieties and Newton polygons.* C. R. Acad. Sci. Paris **312**, Sér. I (1991), 385–389.

[26] F. Oort: *Some questions in algebraic geometry,* preliminary version. Manuscript, June 1995.

[27] F. Oort: *Newton polygons and formal groups: conjectures by Manin and Grothendieck.* Utrecht Preprint 995, January 1997.

[28] F. Oort: *Newton polygon strata in the moduli space of abelian varieties.* In preparation.

[29] G. Prasad: *Strong approximation for semi-simple groups over function fields*. Ann. Math. **105** (1977), 553–572.

[30] Séminaire de Géométrie Algébrique 1962/64 (SGA 3): M. Demazure, A. Grothendieck: *Schémas en groupes*. Vol. I, Lect. Notes Math. 151, Springer Verlag, 1970.

[31] T. Shioda: *Supersingular K3 surfaces*. In: Algebraic Geometry, Copenhagen 1978 (Ed. K. Lønsted). Lect. Notes Math. 732, Springer Verlag, 1979; pp. 564–591.

[32] L. Szpiro: *Propriétés numériques du faisceau dualisant relatif*. Sém. sur les pinceaux de courbes de genre au moins deux. Ed. L. Szpiro. Astérisque 86, Soc. Math. France, 1981; pp. 44–78.

[33] T. Zink: *The display of a formal p-divisible group*. University of Bielefeld, Preprint 98–017, February 1998.

Cycles on the Moduli Space of Abelian Varieties

Gerard van der Geer

Abstract

We first determine the tautological Chow ring of the moduli space $\mathcal{A}_g$ of principally polarized abelian varieties. Then we consider the Ekedahl-Oort stratification on the moduli space $\mathcal{A}_g \otimes \mathbb{F}_p$ in characteristic p and we compute the cycle classes of the strata in this stratification. The formulas for these strata can be seen as a generalization of Deuring's famous formula for the number of supersingular elliptic curves in characteristic p. The formulas for the cycle classes of the strata show that these classes lie in the tautological ring and imply effectivity of certain tautological classes.

1 Introduction

In this paper I present a number of results on cycles on the moduli space $\mathcal{A}_g$ of principally polarized abelian varieties of dimension g. The results on the tautological ring are my own work, the results on the torsion of λ_g and on the cycle classes of the Ekedahl-Oort stratification are joint work with Torsten Ekedahl and some of the results on curves are joint work with Carel Faber. Our results include:

- A description of the tautological subring of the Chow ring of $\mathcal{A}_g$, i.e. of the subring generated by the Chern classes λ_i of the Hodge bundle $\mathbb{E}$.

- A formula for the top Chern class λ_g of the Hodge bundle and a bound for the order of the torsion of this class.

- A description of the Ekedahl-Oort stratification of $\mathcal{A}_g \otimes \mathbb{F}_p$ in terms of degeneracy loci of a map between flag bundles.

- The description of the Chow classes of the strata of this stratification. This includes as special cases formulas for the classes of loci like p-rank $\leq f$ locus or a-number $\geq a$ locus. Such formulas generalize the classical formula of Deuring for the number of supersingular elliptic curves.

- The irreducibility of the locus T_a of abelian varieties of a-number $\geq a$ for $a < g$.

- A computation of this stratification for the Jacobians of hyperelliptic curves of 2-rank 0 in characteristic 2.

- A formula for the class of the supersingular locus for low genera.

Acknowledgments. It is a pleasure for me to acknowledge pleasant cooperation with and help from Torsten Ekedahl and Carel Faber and useful discussions with Hélène Esnault, Frans Oort and Piotr Pragacz. I also would like to thank Kenji Ueno for inviting me to Kyoto in 1996 where I found the time to write the first version of this paper.

2 The Tautological Subring of $\mathcal{A}_g$

Let $\mathcal{A}_g/\mathbb{Z}$ denote the moduli stack of principally polarized abelian varieties of dimension g. This is an irreducible algebraic stack of relative dimension $g(g+1)/2$. We refer to [10] (this volume) for a short introduction. All the fibres $\mathcal{A}_g \otimes \mathbb{F}_p$ and $\mathcal{A}_g \otimes \mathbb{Q}$ are geometrically irreducible. This stack carries a locally free sheaf $\mathbb{E}$ of rank g ("the Hodge bundle") defined by giving for every morphism $S \to \mathcal{A}_g$ a locally free sheaf of rank g which is compatible with pullbacks. It is defined as $s^*\Omega_{A/S}$, where A/S is the principally polarized abelian variety corresponding to $S \to \mathcal{A}_g$ and s is the zero section of A/S. If $\pi : \mathcal{X}_g \to \mathcal{A}_g$ is the universal abelian variety we have $\Omega_{\mathcal{X}_g/\mathcal{A}_g} = \pi^*(\mathbb{E})$.

Let $\tilde{\mathcal{A}}_g$ be a smooth toroidal compactification of $\mathcal{A}_g$. The Hodge bundle can be extended to a locally free sheaf (again denoted by) $\mathbb{E}$ on $\tilde{\mathcal{A}}_g$.

The Chern classes λ_i of the Hodge bundle $\mathbb{E}$ are defined over $\mathbb{Z}$ and for each fibre $\mathcal{A}_g \otimes k$ they give rise to classes λ_i in $CH^*(\mathcal{A}_g \otimes k)$, and in $CH^*(\tilde{\mathcal{A}}_g \otimes k)$. They generate subrings ($\mathbb{Q}$-subalgebras) of $CH^*_{\mathbb{Q}}(\mathcal{A}_g \otimes k)$ and of $CH^*_{\mathbb{Q}}(\tilde{\mathcal{A}}_g \otimes k)$ which we shall call the *tautological subrings*.

We shall describe these tautological subrings, first for $\mathcal{A}_g$ and in the next section for $\tilde{\mathcal{A}}_g$. We derive a relation for the Chern classes λ_i on $\mathcal{A}_g$ in the *Chow ring*.

Theorem 2.1. *The Chern classes λ_i in $CH^*_{\mathbb{Q}}(\mathcal{A}_g)$ in the Chow ring with rational coefficients satisfy the relation*

$$(1 + \lambda_1 + \ldots + \lambda_g)(1 - \lambda_1 + \ldots + (-1)^g \lambda_g) = 1. \tag{1}$$

The idea of the proof is to apply the Grothendieck-Riemann-Roch theorem to the theta divisor on the universal abelian variety $\mathcal{X}_g$ over $\mathcal{A}_g$. We choose this divisor (on a level cover) so that its restriction $s^*(\Theta)$, with s the zero section, is trivial on $\mathcal{A}_g$ and apply Grothendieck-Riemann-Roch to the line bundle $L = O(\Theta)$:

$$
\begin{aligned}
ch(\pi_! L) &= \pi_*(ch(L) \cdot \mathrm{Td}^\vee(\Omega^1_{\mathcal{X}_g/\mathcal{A}_g})) \\
&= \pi_*(ch(L) \cdot \mathrm{Td}^\vee(\pi^*(\mathbb{E}))) \\
&= \pi_*(ch(L)) \cdot \mathrm{Td}^\vee(\mathbb{E}),
\end{aligned}
$$

by the projection formula. Here $\mathrm{Td}^\vee$ is defined for a line bundle with first Chern class α by $\alpha/(e^\alpha - 1)$. Since $R^i \pi_*(L) = 0$ for $i > 0$ it follows that $\pi_!(L)$ is a vector bundle and it is of rank 1 since Θ is a principal polarization. We write $c_1(\pi_!(L)) = \theta$ and find

$$
\sum_{k=0}^{\infty} \frac{\theta^k}{k!} = \pi_*(\sum_{k=0}^{\infty} \frac{\Theta^{g+k}}{(g+k)!}) \cdot \mathrm{Td}^\vee(\mathbb{E}). \tag{2}
$$

Comparison of the terms of degree 1 gives

$$
\theta = -\lambda_1/2 + \pi_*(\Theta^{g+1})/(g+1)!.
$$

Replace now L by $L^{\otimes n}$. The term of degree k in $\pi_*(\sum \Theta^{g+k}/(g+k)!)$ changes by a factor n^{g+k}. But $\pi_!(L^n)$ is a numerical function of degree $\leq n^g$, cf. the arguments of Chai-Faltings on p. 26 of [6]. Or, alternatively, using the Heisenberg group, we find on a suitable cover of $\mathcal{A}_g$

$$
\pi_! L^n = \pi_! L \otimes A,
$$

where A is the standard irreducible representation of dimension n^g of the Heisenberg group. Therefore, up to torsion we have

$$
ch(\pi_! L^n) = n^g ch(\pi_! L).
$$

We thus get

$$
n^g \sum_{k=0}^{\infty} \frac{\theta^k}{k!} = \pi_*(\sum_{k=0}^{\infty} \frac{n^{g+k} \Theta^{g+k}}{(g+k)!}) \cdot \mathrm{Td}(\mathbb{E}^\vee). \tag{2'}
$$

By comparing the coefficients of powers of n we find

$$
\pi_*(\sum_{k=0}^{\infty} \frac{\Theta^{g+k}}{(g+k)!}) = 1 \in CH^*_{\mathbb{Q}}(\mathcal{A}_g). \tag{3}
$$

In particular we get

$$2\theta = -\lambda_1 \qquad \text{(the "key formula")}.$$

Therefore, if we write λ_j as the jth symmetric function of $\alpha_1, \ldots, \alpha_g$ and if we use $\text{Td}(\mathbb{E}^\vee) = \prod(\alpha_i/(e^{\alpha_i} - 1))$ the identity (2) becomes

$$\prod_{i=1}^{g} \frac{e^{\alpha_i/2} - e^{-\alpha_i/2}}{\alpha_i} = 1.$$

and implies $\text{Td}(\mathbb{E} \oplus \mathbb{E}^\vee) = 1$. This is equivalent with $ch_{2k}(\mathbb{E}) = 0$ for $k \geq 1$ and with (1). $\square$

Proposition 2.2. *In the Chow ring $CH_{\mathbb{Q}}^*(\mathcal{A}_g)$ with rational coefficients we have the identity $\lambda_g = 0$.*

Proof. We apply GRR to the structure sheaf $O_{\mathcal{X}}$ of the universal abelian variety $\mathcal{X}$ over the stack $\mathcal{A}_g$. We get

$$ch(\pi_! O_{\mathcal{X}}) = \pi_*(ch(O_{\mathcal{X}}) \cdot \text{Td}^\vee(\Omega^1))$$
$$= \pi_*(1) \text{Td}^\vee(\mathbb{E}),$$

and this gives the relation

$$ch(1 - \mathbb{E}^\vee + \wedge^2 \mathbb{E}^\vee - \ldots + (-1)^g \wedge^g \mathbb{E}^\vee) = \pi_*(1) \text{Td}^\vee(\mathbb{E}) = 0.$$

Now for a vector bundle B of rank r we have in general the relation (see [1])

$$\sum_{j=0}^{r} (-1)^j ch(\wedge^j B^\vee) = c_r(B) \text{Td}(B)^{-1}.$$

This gives: $\lambda_g = 0$ in $CH_{\mathbb{Q}}^*(\mathcal{A}_g)$. $\square$

Let R_g be the quotient of the graded ring $\mathbb{Q}[u_1, \ldots, u_g]$ with generators u_i of degree i by the relation

$$(1 + u_1 + u_2 + \ldots + u_g)(1 - u_1 + u_2 - \ldots + (-1)^g u_g) - 1 = 0.$$

This relation implies (by induction)

$$u_g u_{g-1} \cdots u_{k+1} u_k^2 = 0 \qquad \text{for} \quad k = 1, \ldots, g. \qquad (4)$$

This ring R_g has additive generators

$$u_1^{\epsilon_1} u_2^{\epsilon_2} \ldots u_g^{\epsilon_g} \qquad \text{with} \quad \epsilon_j \in \{0, 1\}.$$

Obviously, we have $R_g/(u_g) \cong R_{g-1}$. It follows from Theorem (2.1) and Proposition (2.2) that the tautological subring of $CH_{\mathbb{Q}}^*(\mathcal{A}_g)$ is a homomorphic image of the ring $R_g/(u_g) \cong R_{g-1}$.

Now consider the moduli space $\mathcal{A}_g \otimes \mathbb{F}_p$. It contains the loci $V_f = V_f(p)$ of abelian varieties with p-rank $\leq f$. Their closures in $\mathcal{A}_g^*$ and in $\tilde{\mathcal{A}}_g \otimes \mathbb{F}_p$ define loci again denoted V_f. These loci have been studied by Koblitz [14], Oort and Norman, cf. [22], [20]. They show that V_0 is complete in $\mathcal{A}_g \otimes \mathbb{F}_p$. We now state a slight generalization of their results. Recall that for a smooth toroidal compactification $\tilde{\mathcal{A}}_g$ one has a morphism $q : \tilde{\mathcal{A}}_g \to \mathcal{A}_g^*$ to the Satake or 'minimal' compactification as defined in [6]. The Satake compactification $\mathcal{A}_g^*$ is a disjoint union

$$\mathcal{A}_g^* = \mathcal{A}_g \sqcup \mathcal{A}_{g-1} \sqcup \ldots \sqcup \mathcal{A}_0.$$

We call the inverse image under q of $\sqcup_{j \leq t} \mathcal{A}_{g-j}$ the moduli space $\tilde{\mathcal{A}}_g^{(t)}$ of rank $\leq t$ degenerations. This space parametrizes semi-stable abelian varieties whose torus-part has rank $\leq t$.

Lemma 2.3. *The subvariety V_f is complete in the moduli space $\tilde{\mathcal{A}}_g^{(f)} \subset \tilde{\mathcal{A}}_g \otimes \mathbb{F}_p$ of rank $\leq f$ degenerations; in particular V_0 is complete in $\mathcal{A}_g$.*

Lemma 2.4. *We have $\lambda_1^{g(g-1)/2+f} \neq 0$ on $\tilde{\mathcal{A}}_g^{(f)}$.*

Proof. Observe that $\det(\mathbb{E})$ is an ample line bundle (its sections are modular forms) and so λ_1 is ample on $\mathcal{A}_g^*$, see [16]. On a complete variety of dimension d the d-th power of an ample divisor is non-zero. $\square$

Theorem 2.5. *The tautological subring of the Chow ring $CH_\mathbb{Q}^*(\mathcal{A}_g)$ generated by the λ_i is isomorphic to R_{g-1}.*

Proof. By the relation $(1 + \lambda_1 + \ldots + \lambda_g)(1 - \lambda_1 + \ldots + (-1)^g \lambda_g) = 1$ we get a quotient ring of R_g. Since moreover $\lambda_g = 0$ we get a quotient of $R_g/(u_g) \cong R_{g-1}$. This ring R_{g-1} is Gorenstein and its top degree elements are proportional. We now use the fact that $\mathcal{A}_g \otimes \mathbb{F}_p$ has a complete subvariety of codimension g, namely the p-rank zero locus. The existence of of a complete subvariety of dimension $g(g-1)/2$ in $\mathcal{A}_g \otimes \mathbb{F}_p$ for every prime p and the ampleness of λ_1 on $\mathcal{A}_g \otimes \mathbb{F}_p$ yield that $\lambda_1^{g(g-1)/2} \neq 0$. This implies that in the quotient of R_{g-1} the one-dimensional socle does not map to zero, hence that the quotient is isomorphic to R_{g-1}. Or more explicitly, consider the set of 2^{g-1} generators of the form

$$\lambda_1^{\epsilon_1} \lambda_2^{\epsilon_2} \cdots \lambda_{g-1}^{\epsilon_{g-1}} \quad \text{with} \quad \epsilon_i \in \{0,1\}.$$

Order these elements λ_ϵ with $\epsilon \in \{0,1\}^{g-1}$ lexicographically. Suppose we have a relation $\sum a_\epsilon \lambda_\epsilon = 0$ in $CH_\mathbb{Q}^*(\mathcal{A}_g)$ and suppose that ϵ' is the 'smallest' exponent. Let ϵ'' be the complementary exponent with $\epsilon' + \epsilon'' = (1, \ldots, 1)$. Then we have

$$\lambda_{\epsilon''}\left(\sum a_\epsilon \lambda_\epsilon\right) = a_{\epsilon'} \lambda_{g-1} \cdots \lambda_1 = 0,$$

and this implies that $a_{\epsilon'} = 0$. By induction the theorem follows. $\square$

Corollary 2.6. *In* $CH^*_{\mathbb{Q}}(\mathcal{A}_g)$ *we have:* $\lambda_1^{g(g-1)/2} \neq 0$ *and* $\lambda_1^{1+g(g-1)/2} = 0$.

Proof. The first statement was given in (2.4). In the ring R_{g-1} all top monomials (of degree $g(g-1)/2$) are proportional, hence $\lambda_1^{g(g-1)/2}$ is a non-zero multiple of the top monomial $\lambda_{g-1}\lambda_{g-2}\cdots\lambda_1$, so $\lambda_1^{g(g-1)/2+1}$ is a non-zero multiple of $\lambda_{g-1}\lambda_{g-2}\cdots\lambda_2\lambda_1^2$, which is zero by (4). $\square$

Corollary 2.7. *Let* $\mathbb{F}$ *be a field. A complete subvariety of* $\mathcal{A}_g \otimes \mathbb{F}$ *has codimension at least* g.

Proof. If Z is a complete subvariety of dimension m then $\lambda_1^m \neq 0$ on Z. Since $g(g-1)/2$ is the highest power of λ_1 which is not zero on $\mathcal{A}_g$ the result follows. $\square$

For a discussion of questions concerning complete subvarieties of $\mathcal{A}_g$ we refer to [23] and [5].

3   The Tautological Ring of $\tilde{\mathcal{A}}_g$

Now we come to the tautological ring of certain compactifications of $\mathcal{A}_g \otimes k$. We first choose a suitable toroidal compactification $\tilde{\mathcal{A}}_g$ as constructed in [6]. We let $\mathcal{A}_g^*$ be the minimal ('Satake') compactification as before. Furthermore, we let $\mathcal{A}_g' = \tilde{\mathcal{A}}_g^{(1)}$ be the moduli space of rank 1-degenerations, i.e. the inverse image of $\mathcal{A}_g \sqcup \mathcal{A}_{g-1} \subset \mathcal{A}_g^*$ under the natural map $q : \tilde{\mathcal{A}}_g \to \mathcal{A}_g^*$. The important point is that the space $\mathcal{A}_g'$ does *not* depend on a choice $\tilde{\mathcal{A}}_g$ of toroidal compactification of $\mathcal{A}_g$. See [19].

Let $\mathcal{G}$ be the 'universal' semi-abelian variety over $\tilde{\mathcal{A}}_g$ with zero section $s : \tilde{\mathcal{A}}_g \to \mathcal{G}$. Then we have a canonical extension defined in a geometric way by $s^*(\mathrm{Lie}\,\mathcal{G})^\vee$ of the Hodge bundle $\mathbb{E}$. By abuse of notation this canonical extension is again denoted by $\mathbb{E}$. Moreover, let $D = \tilde{\mathcal{A}}_g - \mathcal{A}_g$ be the divisor at infinity. Then we have the isomorphism (cf. [6], p. 117):

$$\mathrm{Sym}^2(\mathbb{E}) \cong \Omega^1(\log D).$$

In order to extend the relation (1) for the Chern classes of the Hodge bundle in $CH_{\mathbb{Q}}(\tilde{\mathcal{A}}_g)$ one could try to apply the Grothendieck-Riemann-Roch theorem as we did for $\mathcal{A}_g$. Unfortunately, I was not able to carry this out; for partial results I refer to the next section.

In cohomology with rational coefficients (as opposed to the Chow ring) the relation (1) is known to hold. This follows from the fact that on $\mathcal{A}_g$ we have an exact sequence

$$0 \to \mathbb{E} \longrightarrow \mathcal{H}_{\mathrm{dR}}^1 \longrightarrow \mathbb{E}^\vee \to 0$$

and the Gauss-Manin connection on $\mathcal{H}_{\mathrm{dR}}^1$ and that both the exact sequence and the connection extend to $\tilde{\mathcal{A}}_g$, cf. [6], p. 225.

Now since R_g is a Gorenstein ring with socle in degree $g(g+1)/2$ and since λ_1 is ample on the Zariski-open part $\mathcal{A}_g$ it follows that $\lambda_1^{g(g+1)/2} \neq 0$. Therefore we see that the tautological subring of the cohomology ring $H^*(\tilde{\mathcal{A}}_g(\mathbb{C}), \mathbb{Q})$ is isomorphic to R_g.

Theorem 3.1. *The relation (1) holds for the Chern classes of the Hodge bundle* $\mathbb{E}$ *in* $H^*(\tilde{\mathcal{A}}_g \otimes k, \mathbb{Q})$. *The tautological subring of* $H^*(\tilde{\mathcal{A}}_g \otimes k, \mathbb{Q})$ *is isomorphic to* R_g.

A top monomial u^α in R_g can be written as a multiple $m_\alpha u_1 u_2 \ldots u_g$ with $m_\alpha \in \mathbb{Z}$ using the relation (1). Therefore we find that the degree $\deg \lambda^\alpha$ equals $m_\alpha \deg \lambda_1 \lambda_2 \ldots \lambda_g([\tilde{\mathcal{A}}_g])$.

Let Y_g be the compact dual of the Siegel upper half space $\mathbb{H}_g$. It can be identified with the Grassmann variety of Lagrangian subspaces of a $2g$-dimensional symplectic vector space. The Chow ring and the cohomology ring of Y_g are generated by the Chern classes u_i of the tautological bundle E of rank g on Y_g and are isomorphic to R_g.

The fact that the tautological subring of $H^*_\mathbb{Q}(\tilde{\mathcal{A}}_g) \cong R_g$ gives the following famous 'principle'.

Theorem 3.2. *(Proportionality Principle of Hirzebruch-Mumford) The characteristic numbers of the Hodge bundle are proportional to those of the tautological bundle on* Y_g:

$$\lambda^\alpha([\tilde{\mathcal{A}}_g]) = (-1)^G \frac{1}{2^g} \prod_{k=1}^g \zeta(1-2k) \cdot u^\alpha([Y_g])$$

with $G = g(g+1)/2$.

We have $\lambda^\alpha([\tilde{\mathcal{A}}_g]) = c(g) \times u^\alpha([Y_g])$ for some constant $c(g)$. To evaluate the constant one can compare the Euler number of $\mathcal{A}_g$ (in a suitable sense; actually for $\Omega(\log D) = \mathrm{Sym}^2(\mathbb{E})$) and the Euler number of Y_g. The Euler number of Y_g equals 2^g. We know the Euler number of $\mathcal{A}_g$ by the work of Siegel and Harder:

$$\chi(Sp(2g, \mathbb{Z})) = \frac{\#W_{Sp(\mathbb{R})}}{\#W_{U(\mathbb{R})}} \prod_{j=1}^g \frac{\zeta(1-2j)}{2} = \frac{2^g g!}{g!} \prod_{j=1}^g \frac{\zeta(1-2j)}{2}$$
$$= \zeta(-1)\zeta(-3) \cdots \zeta(-2g+1),$$

where $W_{Sp(\mathbb{R})}$ (resp. $W_{U(\mathbb{R})}$) denotes the Weyl group.

Define a proportionality factor by

$$p(g) = (-1)^G \prod_{j=1}^g \frac{\zeta(1-2j)}{2}.$$

Since $\deg u_1 u_2 \ldots u_g = 1$ we find $\deg \lambda_1 \lambda_2 \cdots \lambda_g = p(g)$. Similarly, using the structure of R_g we have

$$\frac{\lambda_1^G}{\prod_{k=1}^g \zeta(1-2k)} = (-1)^G \frac{u_1^G}{2^g}.$$

and thus get from the relation between u_1^G and $u_1 \cdots u_g$:

$$\lambda_1^G = p(g) G! \prod_{k=1}^g \frac{1}{(2k-1)!!}.$$

This has to be interpreted with care (in the orbifold sense).

Some examples. We have

$$
\begin{aligned}
p(1) &= 1/24 & \deg \lambda_1 &= 1/24, \\
p(2) &= 1/5760 & \deg \lambda_1^3 &= 1/2880, \\
p(3) &= 1/2903040 & \deg \lambda_1^6 &= 1/181440.
\end{aligned}
$$

For the classical Proportionality principle of Hirzebruch we refer to his collected works, [12]. In a letter to Atiyah, Hirzebruch sketches how one obtains a copy of R_g in the cohomology of suitable compact quotients of the Siegel upper half space, cf. [13]. Mumford's extension can be found in [18].

4   Cycle Relations from Grothendieck-Riemann-Roch

In this section we derive cycle relations for the classes λ_i. We begin with λ_g. Since the class λ_g vanishes on $\mathcal{A}_g$ it can be represented in the form $\lambda_g = i_*(x)$ with x a class in $CH_{\mathbb{Q}}^{g-1}(D)$ and $i \colon D \to \tilde{\mathcal{A}}_g$ the embedding of the 'boundary'. By applying Grothendieck-Riemann-Roch to the structure sheaf on the semi-abelian variety over $\tilde{\mathcal{A}}_g$ (as in (1.2)) one can get an expression for x in $CH_{\mathbb{Q}}^*(D)$, as we shall see shortly.

On the other hand we shall show in characteristic p that a multiple of the class λ_g is represented in the Chow ring by the class of a complete subvariety V_0 of $\mathcal{A}_g$. More generally, a multiple of λ_{g-f} is represented by a complete subvariety of $\mathcal{A}_g^{(f)}$, the locus V_f of abelian varieties of p-rank $\leq f$. We thus find interesting representations of the class λ_g. We first consider the Satake compactification.

Lemma 4.1. *The class* $q_*(\lambda_g)$ *in* $CH_{\mathbb{Q}}^g(\mathcal{A}_g^*)$ *is represented by a multiple of the fundamental class of the boundary* $B_g^* = \mathcal{A}_g^* - \mathcal{A}_g$.

Indeed, λ_g is zero on $\mathcal{A}_g$; for dimension reasons $q_* \lambda_g$ is represented by a multiple of the fundamental class of B_g^*.

Proposition 4.2. *The cycle class $[B_g^*]$ of the boundary is the same in the Chow group $CH_{\mathbb{Q}}^g(\mathcal{A}_g^*)$ as a multiple of the class of the image of $\mathcal{A}_{g-1}$ in $\mathcal{A}_g$ under the map $[X] \mapsto [X \times E]$, with E a generic elliptic curve.*

Proof. Consider (for $g > 2$) the space $\mathcal{A}_{g-1,1} \sim \mathcal{A}_{g-1} \times \mathcal{A}_1$ in $\mathcal{A}_g$. Since $\mathcal{A}_1$ is the affine j-line we find a rational equivalence between the cycle class of a generic fibre $\mathcal{A}_{g-1} \times \{j\}$ and a multiple of the fundamental class of the boundary B_g^*. $\square$

Let B_g be the cycle on $\mathcal{A}_g^{(1)}$ defined by the semi-abelian varieties which are trivial extensions

$$1 \to \mathbb{G}_m \to G \to X_{g-1} \to 0,$$

where X_{g-1} is a $(g-1)$-dimensional abelian variety. The cycle $B_g \sim \mathcal{A}_{g-1}$ is of codimension g in $\mathcal{A}_g^{(1)}$ and extends to $\tilde{\mathcal{A}}_g$.

The following proposition describes the class of B_g in cohomology.

Proposition 4.3. *The cohomology class of B_g and of $\mathcal{A}_{g-1} \times \{j\}$ are both equal to $(-1)^g \lambda_g / \zeta(1 - 2g)$.*

Proof. The intersection of the closure of $\mathcal{A}_{g-1,1}$ with the boundary divisor in $\tilde{\mathcal{A}}_g$ is the closure of B_g. Using the rational equivalence of B_g and $\mathcal{A}_{g-1} \times \{j\}$ for generic j and the fact that the class of $\mathcal{A}_{g-1} \times \{j\}$ on $\mathcal{A}_g^*$ is a multiple of $q_* \lambda_g$ the result follows by pullback. To find the multiple we look at characteristic zero, integrate the $Sp(2g, \mathbb{R})$-invariant forms representing the λ_i and use the Proportionality Theorem. $\square$

Examples. We have in cohomology:

$$g = 1 \qquad [B_1] = 12\lambda_1$$
$$g = 2 \qquad [B_2] = 120\lambda_2.$$

Now we try to apply the Grothendieck-Riemann-Roch theorem to the Θ-divisor again on a chosen compactification $\tilde{\mathcal{X}}_g$ of the semi-abelian variety over $\tilde{\mathcal{A}}_g$ as we did for $\mathcal{A}_g$ to obtain cycle relations. We do this on a level cover of $\tilde{\mathcal{A}}_g$ for a line bundle $L = O(T)$ trivialized along the zero section. The relation (1) is equivalent to the identity:

$$\mathrm{Td}^{\vee}(\mathbb{E}) = e^{-\lambda_1/2}.$$

Application of GRR says

$$ch(\pi_!(L^{\otimes n})) = \pi_*(e^{nT} \cdot F) \cdot \mathrm{Td}^{\vee}(\mathbb{E}). \qquad (5)$$

Here we have set $F := \mathrm{Td}^{\vee}(\mathcal{F})^{-1}$, where $\mathcal{F}$ is defined by the exact sequence

$$0 \longrightarrow \Omega_{\tilde{\mathcal{X}}_g/\tilde{\mathcal{A}}_g} \longrightarrow \pi^*(\mathbb{E}) \longrightarrow \mathcal{F} \to 0. \qquad (6)$$

We know by [6] that $\pi^*(\mathbb{E})$ can be identified with the relative logarithmic differentials $\Omega^1(d\log\infty)/\pi^*(\Omega^1(d\log\infty))$ and so $\mathcal{F}$ is concentrated where π is not smooth. The problem is to have a good description of $\pi_!(L^{\otimes n})$.

By applying GRR to the case where $n = 0$, i.e. to the structure sheaf, we obtain the following expression for λ_g, cf. Proposition 2.2.

Proposition 4.4. *We have* $\pi_*(F) = \lambda_g$.

We now state the following result, to be proved in Section 9.

Proposition 4.5. *Let p be a prime. The class $(p-1)(p^2-1)\ldots(p^g-1)\lambda_g \in CH^*_{\mathbb{Q}}(\mathcal{A}_g \otimes \mathbb{F}_p)$ is represented by a complete cycle in $\mathcal{A}_g \otimes \mathbb{F}_p$.*

Corollary 4.6. *In $CH^*_{\mathbb{Q}}(\tilde{\mathcal{A}}_g \otimes \mathbb{F}_p)$ we have the relations $\lambda_g\lambda_{g-1}\ldots\lambda_{k+1}\lambda_k^2 = 0$ for $k = 1,\ldots,g$.*

Proof. By the two preceding propositions we have $\lambda_g^2 = 0$ in $CH^*_{\mathbb{Q}}(\tilde{\mathcal{A}}_g)$. From the relation (1) on $\mathcal{A}_g$ we see that the expressions

$$\lambda_r^2 - 2\lambda_{r-1}\lambda_{r+1} + \ldots + (-1)^{g-r}2\lambda_{2r-g}\lambda_g$$

are boundary classes (living on D). Using induction the relation follows from multiplying these expressions with $\lambda_g\lambda_{g-1}\cdots\lambda_{r+1}$. $\square$

Consider now the moduli space $\mathcal{A}'_g = \tilde{\mathcal{A}}_g^{(1)}$ of rank ≤ 1 degenerations. This is a *canonical* partial compactification of $\mathcal{A}_g$. If we want a (complete) compactification then there is not really a unique one, but we must make choices. Let D^0 be the closed subset corresponding to rank 1 degenerations. The divisor D^0 has a morphism $\phi : D^0 \to \mathcal{A}_{g-1}$ which exhibits D^0 as the universal abelian variety over $\mathcal{A}_{g-1}$. The fibre over $x \in \mathcal{A}_{g-1}$ is the dual $\hat{X}_{g-1}$ of the abelian variety X corresponding to x. The 'universal' semi-abelian variety G over $\hat{\mathcal{X}}_{g-1}$ is the $\mathbb{G}_m$-bundle obtained from the Poincaré bundle $P \to \mathcal{X}_{g-1} \times \hat{\mathcal{X}}_{g-1}$ by deleting the zero-section. We have the maps

$$G = P - \{(0)\} \to \mathcal{X}_{g-1} \times \hat{\mathcal{X}}_{g-1} \to \mathcal{A}_{g-1}.$$

By considering the cotangent bundle to G at the zero section t of $G \to \hat{X}_{g-1}$ we get an exact sequence

$$0 \to q^*\mathbb{E}_{g-1} \to \mathbb{E}_{g|D^0} \to U \to 0,$$

with U the pullback of a line bundle on $\mathcal{A}^*_{g-1}$. Now U is trivial since the restriction of $\mathbb{E}_g$ to B_g is a direct sum of $\mathbb{E}_{g-1}$ and a trivial line bundle.

Consider $\mathcal{X}'$, the compactified family of semi-abelian varieties over the moduli stack $\mathcal{A}'_g$ of rank ≤ 1 degenerations. A semi-abelian variety which is a $\mathbb{G}_m$-bundle over an abelian variety X_{g-1} is compactified by taking the $\mathbb{P}^1$-bundle associated to the $\mathbb{G}_m$-bundle and then identifying the 0- and the ∞-section under a shift $\xi \in X_{g-1}$. The image of the zero section of the $\mathbb{P}^1$-bundle maps to a codimension 2 cycle Δ, the locus of singular points of the fibres of $\mathcal{X}'_g \to \mathcal{A}'_g$. We have

$$\Delta \cong \mathcal{X}_{g-1} \times_{\mathcal{A}_{g-1}} \hat{\mathcal{X}}_{g-1}.$$

We analyze $\Omega^1 = \Omega_{\mathcal{X}'_g/\mathcal{A}'_g}$. In the exact sequence

$$0 \to \Omega^1 \to \pi^*(\mathbb{E}) \to \mathcal{F} \to 0, \qquad (6')$$

the sheaf $\mathcal{F}$ has support on Δ, the codimension 2 cycle. Let u be a fibre coordinate on a $\mathbb{G}_m$-bundle over $\mathcal{X}_{g-1}$. A section of $\pi^*(\mathbb{E})$ is given locally by du/u. Pull a section back to the $\mathbb{P}^1$-bundle and take the residue along the 0- and ∞-section. The residue map yields an isomorphism on $\mathcal{A}'_g$

$$\mathcal{F} \cong O_{\tilde{\Delta}},$$

where $\tilde{\Delta}$ is the etale double cover of Δ corresponding to choosing the branches (0 and ∞). Working out now the expression for F on $\mathcal{A}'_g$ one can show (cf. [3]):

Theorem 4.7. *The Chow class of B_g and that of $A_{g-1} \times \{j\}$ in $CH^g_{\mathbb{Q}}(\mathcal{A}'_g)$ are both equal to $(-1)^g \lambda_g / \zeta(1-2g)$.*

The GRR theorem for the case $n = 1$ yields the identity:

$$ch(\pi_! L) = \pi_*(e^T F) \mathrm{Td}^{\vee}(\mathbb{E}).$$

The expression $\pi_! L$ is given by a line bundle. So to calculate it we may work on $\mathcal{A}'_g$ (instead of $\tilde{\mathcal{A}}_g$). We write

$$ch(\pi_!(L^{\otimes n})) = 1 + \theta_1^{(n)} + \theta_2^{(n)} + \dots$$

and set $\theta_1^{(1)} = \theta$. In equation (5) we compare terms of codimension 1:

$$\theta_1^{(n)} = -\frac{\lambda_1}{2} \cdot \pi_*[e^{nT} \cdot \mathrm{Td}^{\vee}(O_\Delta)^{-1})]_0 + \pi_*[e^{nT} \cdot \mathrm{Td}^{\vee}(O_\Delta)^{-1})]_1$$

Lemma 4.8. *In the Chow group $CH^1_{\mathbb{Q}}(\mathcal{A}'_g)$ we have $\theta = -\lambda_1/2 + \delta/8$, where δ is the class of the 'boundary' D.*

Proof. First we do the case $g = 1$. We have

$$1 + \theta = (\pi_*(T + T^2/2) + \pi_*(\Delta/12))(1 - \lambda_1/2).$$

Let S be the zero-section. By Kodaira's results on elliptic surfaces we have $S^2 = -\lambda_1$; so the normalized T is $T = S + \pi^*(\lambda_1)$ with $T^2 = S \cdot \pi^*(\lambda_1)$, i.e. $\pi_*(T^2/2) = \lambda_1/2$. We find $\theta = -\lambda_1/2 + \lambda_1/2 + \delta/12$. Using the relation $12\lambda_1 = \delta$ we can rewrite this as $\theta = -\lambda_1/2 + \delta/8$. For general g we have a priori $\theta = -\lambda_1/2 + a\delta$ for some $a \in \mathbb{Q}$. Restriction to the space of products of elliptic curves $\mathcal{A}_{1,\dots,1}$ gives $a = 1/8$. $\square$

As a corollary we find for the codimension 1 terms of $ch(\pi_!(L^{\otimes n}))$ the identity

$$\theta_1^{(n)} = n^g(-\lambda_1/2) + (n^{g+1} + 2n^{g-1})\delta/24.$$

For a description of the Chow rings of $\tilde{\mathcal{A}}_g(\mathbb{C})$ for $g = 1,2,3$ see [8].

In cohomology the relation (1) holds. Substitution of this result in the GRR formula gives the following identity.

Theorem 4.9. *We have $\pi_*(e^T F) = e^{\delta/8}$ in $H^*(\tilde{\mathcal{A}}_g, \mathbb{Q})$.*

5   On the Torsion of the Class λ_g

By Proposition (2.2) we know that the top Chern class λ_g of $\mathbb{E}$ vanishes in the rational Chow group $CH_{\mathbb{Q}}^g(\mathcal{A}_g)$. So λ_g is a torsion class on $\mathcal{A}_g$. Mumford proved in [17] that the order of λ_1 in $CH^1(\mathcal{A}_1)$ is 12. In [3] we shall prove the following bound for the order of the torsion class λ_g.

Definition. Let n_g be the greatest common divisor of all $p^{2g} - 1$ where p runs through all primes greater than $2g + 1$.

We need a little lemma.

Lemma 5.1. *We have*

$$\prod_{i=1}^{g} n_i = \prod_{p\,prime} \left(\left[\frac{2gp}{p-1} \right]! \right)_p \quad (= \text{multiple of denominator of } p(g)\,).$$

Theorem 5.2. *On $\mathcal{A}_g$ we have:* $(g-1)! \left(\prod_{i=1}^{g} n_i \right) \lambda_g = 0$.

Example i) For $g = 1$ we get $24\lambda_1 = 0$ which is off by a factor 2. ii) For $g = 2$ we get $24 \cdot (16 \cdot 3 \cdot 5)\lambda_2 = 0$. iii) For $g = 3$ we get $2 \cdot 24 \cdot (16 \cdot 3 \cdot 5) \cdot (8 \cdot 9 \cdot 7)\lambda_3 = 0$. We refer to [3] for other cycle relations in the Chow ring of $\tilde{\mathcal{A}}_g$.

6   The Ekedahl-Oort Stratification in Positive Characteristic

Ekedahl and Oort introduced a stratification of the moduli space $\mathcal{A}_g \otimes \mathbb{F}_p$, cf. Oort's paper [24] in this volume. It is defined by analyzing for abelian varieties X the action of Frobenius and Verschiebung on the group scheme $X[p]$, the kernel of multiplication by p. The strata include the well-known loci V_f of abelian varieties of p-rank $\leq f$ (for $0 \leq f \leq g$) and the loci T_a of abelian varieties with a-number $\geq a$ (for $0 \leq a \leq g$). This stratification is very useful for understanding the structure of $\mathcal{A}_g$ both in characteristic p as well as in characteristic 0.

We shall describe these strata in a somewhat different way using the Hodge bundle $\mathbb{E}$. We then can apply theorems of Porteous type to calculate the cycle classes of these loci. Such Porteous type formulas are obtained by applying results of Fulton and of Pragacz on degeneracy maps between symplectic bundles. The cycle classes all lie in the tautological ring. We get explicit formulas for the loci V_f and T_a generalizing Deuring's formula for the number of supersingular elliptic curves and the formula for the number of superspecial abelian varieties (those with $a = g$).

In the following we shall study the moduli space $\mathcal{A}_g \otimes \mathbb{F}_p$ of principally polarized abelian varieties in characteristic p. For simplicity we shall write $\mathcal{A}_g$ instead of $\mathcal{A}_g \otimes \mathbb{F}_p$.

Recall the *canonical filtration* on the de Rham cohomology of a principally polarized abelian variety X as defined by Ekedahl and Oort. They used the group scheme $X[p]$ (kernel of multiplication by p) to define it. We use de Rham-cohomology.

We write $G = H^1_{dR}(X)$ on which we have a σ-homomorphism F and a σ^{-1}-homomorphism V. For a moment we shall ignore the $\sigma^{\pm}$-linearity. We have $FV = VF = 0$. The $2g$-dimensional space is provided with a symplectic form $\langle\,,\rangle$ and F and V are adjoints: $\langle Vg, g' \rangle = \langle g, Fg' \rangle$. For any subspace H of G we have $(VH)^\perp = F^{-1}(H^\perp)$. The spaces $V(G) = F^{-1}(0)$ and $F(G) = V^{-1}(0)$ are maximally isotropic subspaces of dimension g.

To construct the canonical filtration one starts with $0 \subset G$ and constructs finer filtrations by adding the images of V and the orthogonal complements of the spaces present. This process stops. The filtration obtained is stable under V and under $\perp$, hence under F^{-1} as well and is called the *canonical filtration*. The canonical filtration

$$0 \subset C_1 \subset \ldots \subset C_r \subset C_{r+1} \subset \ldots \subset C_{2r}$$

satisfies $C_r = V(C_{2r})$ and $C_{r-i}^\perp = C_{r+i}$. This filtration can be refined to a so-called *final filtration* by choosing a V- and $\perp$-stable filtration of length $2g$ which refines the canonical one. In general, there is no unique choice for a final filtration. We thus get a filtration

$$0 \subset G_1 \subset \ldots \subset G_g \subset G_{g+1} \subset \ldots \subset G_{2g}$$

which satisfies $V(G_{2g}) = G_g$, $G_{g-i}^\perp = G_{g+i}$ and now also $\dim(G_i) = i$. The associated *final type* is the increasing and surjective map

$$\nu : \{0,1,2,\ldots,2g\} \to \{0,1,2,\ldots,g\}$$

satisfying

$$\nu(2g - i) = \nu(i) - i + g \qquad \text{for} \quad 0 \le i \le g \tag{9}$$

obtained by $\nu(i) = \dim(V(G_i))$ and $\nu(0) = 0$. The *canonical type* is the restriction of ν to the integers which arise as dimensions of the C_i. Although the final filtration is not unique, the final type is. (The dimensions between two steps of the canonical filtration either remain constant or grow each step by 1.)

Example. Let X be an abelian variety with p-rank f and $a(X) = 1$ (equivalently, on G_g the operator V has rank $g - 1$ and semi-simple rank $g - f$). Then the canonical type is given by the numbers $\{\text{rank}(C_i)\} =$

$$\{0 < f < f + 1 < \ldots < g - 1 < g < g + 1 < \ldots < 2g - f - 1 < 2g - f < 2g\}$$

and

$$\nu(f) = f, \; \nu(f+1) = f,$$
$$\nu(f+2) = f+1, \ldots, \nu(g) = g-1, \ldots, \nu(2g-f-1) = g-1,$$
$$\nu(2g-f) = g, \; \nu(2g) = g.$$

It is not difficult to see that there is a bijection between the set of final types and the set of canonical types and we have 2^g of them. Note that in view of (9) the function ν is determined by its restriction to $\{1,2,\ldots,g\}$. This restriction is again denoted by ν.

The Ekedahl-Oort stratification of $\mathcal{A}_g$ is obtained by looking for each geometric point of $\mathcal{A}_g$ what the canonical type (or final type) is. The set Z_ν of all abelian varieties which have given final type ν is locally closed and these Z_ν define a stratification, cf. [24].

I prefer to describe the combinatorial datum of a final type ν by a *partition* $\mu = \{g \geq \mu_1 > \mu_2 > \ldots > \mu_r\}$ as follows:

$$\mu_j = \#\{i : 1 \leq i \leq g, \nu(i) \leq i-j\};$$

equivalently, we can visualize it by the associated Young-type diagram with μ_j squares in the j-th layer (i.e. by putting a stack of $i - \nu(i)$ squares in position i):

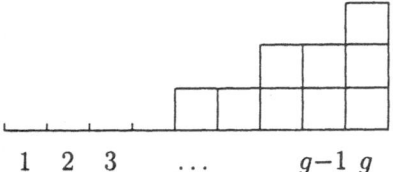

$$1 \quad 2 \quad 3 \qquad \ldots \qquad g{-}1 \;\; g$$

This example corresponds to $\{\nu(i) : i = 1,...,g\} = \{1,2,\ldots,g-5,g-5,g-4,g-4,g-3,g-3\}$ and to $\mu = \{5,3,1\}$.

Definition. We call a partition μ *admissible* if $g \geq \mu_1 > \mu_2 > \ldots > \mu_r > 0$. We call the function $\nu: \{1,\ldots,g\} \to \{1,\ldots,g\}$ *admissible* if

$$0 \leq \nu(i) \leq \nu(i+1) \leq \nu(i)+1 \leq g+1$$

for $1 \leq i \leq g$. The number $|\mu| := \sum_i \mu_i$ is called the *area of the diagram*.

The notions 'admissible' diagram (or partition) and 'admissible function' ν and final type are all equivalent. The set of admissible diagrams carries a partial ordering in the obvious way: $\mu \geq \mu'$ if $\mu_i \geq \mu_i'$ for all i.

7   The Ekedahl-Oort Strata as Degeneracy Loci

We shall now give another approach to these strata by defining them globally on a flag space over $\mathcal{A}_g$. Our starting point is the observation that if we define for a final filtration E_i $(1 \leq i \leq g)$ another filtration by $F_g = \ker(V) = V^{-1}(0)$, by $F_{g+i} = V^{-1}(E_i)$ for $i = 1,...,g$ and by $F_{g-i} = F_{g+i}^{\perp}$ then we have

$$V(E_i) \subseteq E_{\nu(i)} \iff E_i \subseteq V^{-1}(E_{\nu(i)}) = F_{g+\nu(i)} \iff \dim(E_i \cap F_{g+\nu(i)}) \geq i. \quad (10)$$

Working now globally, we let S be a scheme in characteristic p and let $\mathcal{X} \to S$ be an abelian variety over S with principal polarization. Then we consider the de Rham cohomology sheaf $\mathcal{H}^1_{dR}(\mathcal{X}/S)$. It is defined as the hyper-direct image $\mathcal{R}^1 \pi_*(O_\mathcal{X} \to \Omega^1_{\mathcal{X}/S})$. It is a locally free sheaf of rank $2g$ on S equipped with a non-degenerate alternating form (cf. [21]).

$$\langle , \rangle \colon \mathcal{H}^1_{dR}(\mathcal{X}/S) \times \mathcal{H}^1_{dR}(\mathcal{X}/S) \to O_S.$$

Indeed, the polarization (locally in the étale topology given by a relatively ample line bundle on $\mathcal{X}/S$) provides us with a symmetric homomorphism $\rho \colon \mathcal{X} \to \hat{\mathcal{X}}$ and the Poincaré bundle defines a perfect pairing between $\mathcal{H}^1_{dR}(\mathcal{X}/S)$ and $\mathcal{H}^1_{dR}(\hat{\mathcal{X}}/S)$. Moreover, we have an exact sequence of locally free sheaves

$$0 \to \pi_*(\Omega^1) \to \mathcal{H}^1_{dR}(\mathcal{X}/S) \to R^1\pi_* O_\mathcal{X} \to 0.$$

We shall write $\mathbb{H}$ for the sheaf $\mathcal{H}^1_{dR}(\mathcal{X}_g/\mathcal{A}_g)$ and $\mathcal{X}$ for $\mathcal{X}_g$. We thus have an exact sequence

$$0 \to \mathbb{E} \to \mathbb{H} \to \mathbb{E}^\vee \to 0.$$

The relative Frobenius $F \colon \mathcal{X} \to \mathcal{X}^{(p)}$ and the Verschiebung $V \colon \mathcal{X}^{(p)} \to \mathcal{X}$ satisfy $F \cdot V = p \cdot \mathrm{id}_{\mathcal{X}^{(p)}}$ and $V \cdot F = p \cdot \mathrm{id}_\mathcal{X}$ and they induce maps in cohomology, again denoted by F and V:

$$F \colon \mathbb{H}^{(p)} \to \mathbb{H} \qquad \text{and} \qquad V \colon \mathbb{H} \to \mathbb{H}^{(p)}.$$

Of course, we have $FV = 0$ and $VF = 0$ and F and V are adjoints. This implies that $\mathrm{Im}(F) = \ker(V)$ and $\mathrm{Im}(V) = \ker(F)$ are maximally isotropic subbundles of $\mathbb{H}$ and $\mathbb{H}^{(p)}$. Moreover, since $dF = 0$ on $\mathrm{Lie}(\mathcal{X})$ it follows that $F = 0$ on $\mathbb{E}$ and thus $\mathrm{Im}(V) = \mathbb{E}^{(p)}$. Verschiebung thus provides us with a bundle map (again denoted by V): $V \colon \mathbb{H} \to \mathbb{E}^{(p)}$. Consider the space of symplectic flags $\mathcal{F} = \mathrm{Flag}(\mathbb{H})$ on the bundle $\mathbb{H}$. This space is fibred by the spaces $\mathcal{F}^{(i)}$ of partial flags

$$\mathbb{E}_i \subsetneq \mathbb{E}_{i+1} \subsetneq \ldots \subsetneq \mathbb{E}_g.$$

So $\mathcal{F} = \mathcal{F}^{(1)} = \mathrm{Flag}(\mathbb{H})$ and $\mathcal{F}^{(g)} = \mathcal{A}_g$ and there are natural maps

$$\pi_{i,i+1} \colon \mathcal{F}^{(i)} \to \mathcal{F}^{(i+1)}.$$

The fibres are Grassmann varieties of dimension i. The space $\mathcal{F}^i$ is equipped with a universal flag. On $\mathcal{F}$ the Chern classes of the bundle $\mathbb{E}$ decompose into their roots:

$$\lambda_i = \sigma_i(\ell_1, \ldots, \ell_g) \qquad \text{with} \qquad \ell_i = c_1(\mathbb{E}_i/\mathbb{E}_{i-1}).$$

Given an arbitrary flag of subbundles

$$0 \subsetneq \mathbb{E}_1 \subsetneq \ldots \subsetneq \mathbb{E}_g = \mathbb{E} \tag{11}$$

with $\operatorname{rank}(\mathbb{E}_i) = i$ we can extend this to a symplectic filtration on $\mathbb{H}$ by putting

$$\mathbb{E}_{g+i} = (\mathbb{E}_{g-i})^{\perp}.$$

By base change we can transport this filtration to $\mathbb{H}^{(p)}$.

We introduce a second filtration by starting with the isotropic subbundle

$$\mathbb{F}_g = \ker(V) = V^{-1}(0) \subset \mathbb{H}$$

and continuing with

$$\mathbb{F}_{g+i} = V^{-1}(\mathbb{E}_i^{(p)}) \qquad \text{for} \quad 1 \leq i \leq g.$$

We extend it to a symplectic filtration by setting $\mathbb{F}_{g-i} = (\mathbb{F}_{g+i})^{\perp}$. We thus have two filtrations $\mathbb{E}_{\bullet}$ and $\mathbb{F}_{\bullet}$ on $\mathbb{H}$.

Example. i) Let X be an ordinary abelian variety. Then $\operatorname{Lie}(X) = \operatorname{Lie}(\mu)$ with μ the multiplicative subgroup scheme of $X[p]$ of rank p^g. It follows that V is invertible on $\operatorname{Lie}(X)^{\vee} = \omega(X)$, i.e. $\mathbb{F}_g \cap \mathbb{E}_g = (0)$.

ii) If X is a superspecial abelian variety (i.e. equivalently, $a = g$ or X without its polarization is a product of supersingular elliptic curves) then $V = 0$ on $\omega(X)$ so that $\operatorname{rk}(\mathbb{E}_i \cap \mathbb{F}_g) = i$.

These two (extreme) examples show that the respective position of the two filtrations $\mathbb{E}_{\bullet}$ and $\mathbb{F}_{\bullet}$ for an abelian variety X gives information on the structure of the kernel of multiplication by p on X. These respective positions are encoded by a combinatorial datum, e.g. ν or to be more precise, by an element of a Weyl group. We shall associate strata to such data.

To either ν or μ corresponds an element of the Weyl group of the symplectic group. The Weyl group W_g of type C_g in Cartan's terminology is isomorphic to the semi-direct product $S_g \ltimes (\mathbb{Z}/2\mathbb{Z})^g$, where S_g acts on $(\mathbb{Z}/2\mathbb{Z})^g$ by permuting the g factors. Another description of this group is as the subgroup of S_{2g} of elements which map any symmetric 2-element subset of the form $\{i, 2g+1-i\}$ of $\{1, \ldots, 2g\}$ to a subset of the same type:

$$W_g = \{\sigma \in S_{2g} : \sigma(i) + \sigma(2g+1-i) = 2g+1 \text{ for } i = 1, \ldots, g\}.$$

An element in this Weyl group has a *length* and a *codimension*:

$$\ell(w) = \#\{i < j : w(i) > w(j)\} + \#\{i < j : w(i) + w(j) > 2g+1\}$$

and

$$\operatorname{codim}(w) = \#\{i < j : w(i) < w(j)\} + \#\{i < j : w(i) + w(j) < 2g+1\}.$$

We have the equality

$$\ell(w) + \operatorname{codim}(w) = g^2.$$

To a function ν we associate the following element of the Weyl group, a permutation of $\{1,2,\dots,2g\}$: let

$$S = \{i_1,i_2,\dots\} = \{1 \le i \le g : \nu(i) = \nu(i-1)\}$$

with $i_1 < i_2 < \dots$ given in increasing order. Let

$$S^c = \{j_1,j_2,\dots\}$$

be the elements of $\{1,2,\dots,g\}$ not in S, in increasing order. Then one gets the permutation of S_{2g} defined by ν by writing $g+k$ at position j_k for $k = 1,2,\dots$ and k at position i_k for $k = 1,2,\dots$. We finish off by putting $2g+1-k$ at position $2g+1-i$ if k is written at position i. We obtain a sequence s which is a permutation of $\{1,2,\dots,2g\}$.

Alternatively, using diagrams, we can describe the element $w = w_\mu$ as follows. Let t_i be the operator on the set of diagrams which is 'remove the top box of the i-th column'. We let the t_i act from the right. Given an admissible diagram μ we consider the complementary diagram

$$\mu^c = \{g,g-1,\dots,1\} - \mu$$

which is also an admissible diagram (partition). We can successively apply operators t_i to it such that after every step we obtain an admissible diagram and such that after $|\mu^c|$ steps we obtain the empty diagram $(\mu^c) \cdot t_{i_1} \cdots t_{i_\ell} = \emptyset$. We remove first the top layer, then the next one and so on. For $1 \le i < g$ let $s_i \in S_{2g}$ be the permutation $(i,i+1)(2g-i,2g+1-i)$ and let $s_g = (g,g+1) \in S_{2g}$. Now associate to the diagram μ the element of the Weyl group:

$$\mu \mapsto w_\mu = s_{i_1} \cdots s_{i_\ell}.$$

Each diagram thus yields an element in the Weyl group. The admissible partitions yield 2^g elements of the $2^g g!$ elements of W_g.

Example. $g = 3$

μ	ν	$[1,2,3,4,5,6] \mapsto$	ℓ	w_ℓ
$\emptyset$	$\{1,2,3\}$	$[4,5,6,1,2,3]$	6	$s_3 s_2 s_3 s_1 s_2 s_3$
$\{1\}$	$\{1,2,2\}$	$[4,5,1,6,2,3]$	5	$s_2 s_3 s_1 s_2 s_3$
$\{2\}$	$\{1,1,2\}$	$[4,1,5,2,6,3]$	4	$s_3 s_1 s_2 s_3$
$\{3\}$	$\{0,1,2\}$	$[1,4,5,2,3,6]$	3	$s_3 s_2 s_3$
$\{2,1\}$	$\{1,1,1\}$	$[4,1,2,5,6,3]$	3	$s_1 s_2 s_3$
$\{3,1\}$	$\{0,1,1\}$	$[1,4,2,5,3,6]$	2	$s_2 s_3$
$\{3,2\}$	$\{0,0,1\}$	$[1,2,4,3,5,6]$	1	s_3
$\{3,2,1\}$	$\{0,0,0\}$	$[1,2,3,4,5,6]$	0	1

We can associate to the map $V : \mathbb{H} \to \mathbb{E}^{(p)}$ and an element w a degeneracy locus $\mathcal{U}_w$ and in particular to an admissible diagram μ a degeneracy locus $\mathcal{U}_\mu = \mathcal{U}_{w_\mu}$ in $\mathcal{F} = \mathrm{Flag}(\mathbb{H})$. Intuitively, $\mathcal{U}_w$ is defined as the locus of points x such that at x we have

$$\dim(\mathbb{E}_i \cap \mathbb{F}_j) \geq \#\{a \leq i : w(a) \leq j\} \quad \text{for all} \quad 1 \leq i, j \leq g$$

or equivalently, that $\ker(V) \cap \mathbb{E}_i \geq i - \nu(i)$ (= the number of squares in the diagram μ in position i). For the precise definition of degeneracy loci of bundle maps we refer to Fulton [7]. Note that for the admissible diagrams we do not use the full filtration of $\mathbb{F}$, but only $\mathbb{F}_g$. Note that our strata (the degeneracy loci) are closed.

8   The Degeneracy Locus $\mathcal{U}_\emptyset$ and $\Gamma_0(p)$

We first look at the case of the empty diagram $\mu = \emptyset$, or equivalently, that $\nu = \{1, 2, 3, \dots, g\}$. The degeneracy conditions say that (cf. (10))

$$V(\mathbb{E}_i) \subseteq \mathbb{E}_i^{(p)} \qquad i = 1, \dots, g.$$

i.e. we are looking at the space $\mathcal{U}_{\{\emptyset\}}$ of symplectic filtrations on $\mathbb{E}$ which are compatible with the action of V. The codimension of this space in $\mathcal{F}$ is $g(g-1)/2$, hence $\dim(U_{\{\emptyset\}}) = g(g+1)/2 = \dim(\mathcal{A}_g)$. We have a finite map $\pi : U_{\{\emptyset\}} \to \mathcal{A}_g$ of degree

$$\deg(\pi) = (1+p)(1+p+p^2) \dots (1+p+p^2+\dots+p^{g-1}).$$

This space $\mathcal{U} = U_{\{\emptyset\}}$ can be seen as a component of the moduli space $\Gamma_0(p)$. Indeed, a filtration of the subgroup scheme $X[p]$ corresponds to a filtration on H^1_{dR}. Fix g and let $\Gamma_0(p)'$ be the functor that associates to a scheme S the set of isomorphism classes of principally polarized abelian schemes X over S plus a V-stable symplectic filtration on $X[p]$.

Proposition 8.1. *The degeneracy cycle $\mathcal{U}_\emptyset$ is the algebraic stack representing the functor $\Gamma_0(p)'$. The stack $\mathcal{U}_\emptyset$ is fibred by finite morphisms*

$$\mathcal{U}_\emptyset = \mathcal{U}^{(1)} \xrightarrow{\pi_{1,2}} \mathcal{U}^{(2)} \xrightarrow{\pi_{2,3}} \dots \xrightarrow{\pi_{g-1,g}} \mathcal{U}^{(g)} = \mathcal{A}_g.$$

with $\deg(\pi_{i,i+1}) = 1 + p + \dots + p^i$. It contains the degeneracy loci $\mathcal{U}_\mu$ for all 2^g partitions μ.

9 The Cycle Classes of the Strata

In this section we try to compute the cycle classes of the stratification on $\mathcal{A}_g \otimes \mathbb{F}_p$ and show that these classes lie in the tautological ring.

The stacks $\mathcal{U}^{(i)}$ come with universal (partial) flags $\mathbb{E}_i \subsetneq \ldots \subsetneq \mathbb{E}_g$. We denote by $\lambda_j(i) = c_j(\mathbb{E}_i)$ the Chern class in $CH^j_{\mathbb{Q}}(\mathcal{U}^{(i)})$. We also have tautological quotient bundles $L_i = \mathbb{E}_i/\mathbb{E}_{i-1}$. We denote by l_i the Chern class of L_i. We have $(\lambda_j)(i) = \sigma_j(l_1, \ldots, l_i)$, the j-th elementary symmetric function of the $l_1, \ldots, l_i$.

Next we look at the cases where μ is a partition of the form $\mu = \{\mu_1 = g - f\}$. The corresponding degeneracy loci classify the loci of p-rank $\leq f$. It is well-known by Koblitz (cf. [14], [20]) that the codimension of V_f in $\mathcal{A}_g$ equals $g - f$. These loci admit the following explicit description. The pullback of V_{g-1} to $\mathcal{U}_\emptyset$ consists of g components, say $Z_1, \ldots, Z_g$. An abelian variety of p-rank $g-1$ and $a = 1$ has a unique subgroup scheme α_p. The index i of Z_i indicates where for the generic point of Z_i the α_p can be found: $\alpha_p \subset G_i$, $\alpha_p \not\subset G_{i-1}$.

Then the pullback of V_f consists of

$$\pi^{-1}(V_f) = \sum_{\#S = g-f} Z^S,$$

where for a subset $S \subset \{1, \ldots, g\}$ the cycle Z^S is defined as

$$Z^S = \cap_{i \in S} Z_i.$$

Then one sees easily that the cycle V_{g-f} on $\mathcal{A}_g$ is obtained as the pushforward of $Z_g \cap \ldots \cap Z_{g-f+1}$.

Lemma 9.1. *The cycle class of Z_i is equal to $(p-1)\ell_i$.*

Proof. It is described as the locus where $\det(V) : L_i \to L_i^{(p)}$ vanishes. By viewing $\det(V)$ as a section of $L_i^{(p)} \otimes L_i^{-1}$ the result follows. $\square$

We find after a calculation:

$$\pi^*([V_f]) = (p-1)^{g-f} \lambda_{g-f}$$

on $\mathcal{U}_\emptyset$. Using the pushforward of the $\pi_{i,i+1}$ on the classes ℓ_i and extending the result to a compactification $\tilde{\mathcal{A}}_g$ we get:

Theorem 9.2. *The cycle class $[V_f]$ of V_f, the p-rank $\leq f$ locus, in the Chow ring $CH^*_{\mathbb{Q}}(\tilde{\mathcal{A}}_g)$, is given by $[V_f] = (p-1)(p^2-1)\ldots(p^{g-f}-1)\lambda_{g-f}$.*

By applying this formula in the simplest case where $g = 1$ and $f = 0$ we get a classical result.

Corollary 9.3. *(Deuring Mass Formula) Let $g = 1$. We have*

$$\sum_E \frac{1}{\#\mathrm{Aut}(E)} = \frac{p-1}{24},$$

where the sum is over the isomorphism classes over $\overline{\mathbb{F}}_p$ of supersingular elliptic curves.

Proof. The formula gives $(p-1)\lambda_1$ and λ_1 equals $\delta/12$. The class of δ is equivalent to $1/2$ times the class of a 'physical' point of the j-line because the degenerate elliptic curve corresponding to δ has 2 automorphisms. $\square$

Another case where we can find an explicit formula is that of the locus T_a. This corresponds to $\mu = \{a, a-1, \ldots, 2, 1\}$. But here we can work directly on $\mathcal{A}_g$. The locus T_a on $\mathcal{A}_g$ may be defined as the locus

$$\{x \in \mathcal{A}_g : \mathrm{rank}(V)|_{\mathbb{E}_g} \leq g - a\}.$$

We have $T_{a+1} \subset T_a$ and $\dim(T_g) = 0$.

Pragacz and Ratajski, cf. [26], have developed formulas for the degeneracy locus for the rank of a self-adjoint bundle map of symplectic bundles globalizing the results in isotropic Schubert calculus from [25]. Before we apply their result to our case we have to introduce some notation.

Define for a vector bundle A with Chern classes a_i the expression

$$Q_{ij}(A) := a_i a_j + 2 \sum_{k=1}^{j} (-1)^k a_{i+k} a_{j-k} \quad \text{for} \quad i > j.$$

For a admissible partition $\beta = (\beta_1, \ldots, \beta_r)$ (with r even, β_r may be zero) we set

$$Q_\beta = \mathrm{Pfaffian}(x_{ij}),$$

where the (x_{ij}) is an anti-symmetric matrix with $x_{ij} = Q_{\beta_i, \beta_j}$. Applying the Pragacz-Ratajski formula to our situation gives the following result:

Theorem 9.4. *The cycle class $[T_a]$ of the reduced locus T_a of abelian varieties with a-number $\geq a$ is given by*

$$\sum Q_\beta(\mathbb{E}^{(p)}) \cdot Q_{\rho(a)-\beta}(\mathbb{E}^*),$$

where the sum is over the admissible partitions β contained in the partition $\rho(a) = (a, a-1, a-2, \ldots, 1)$.

Example:

$$[T_1] = p\lambda_1 - \lambda_1$$
$$[T_2] = (p-1)(p^2+1)(\lambda_1\lambda_2) - (p^3-1)2\lambda_3$$
$$\cdots$$
$$[T_g] = (p-1)(p^2+1)\ldots(p^g+(-1)^g)\lambda_1\lambda_2\cdots\lambda_g.$$

As a corollary we find a classical result of Ekedahl (cf. [2]) on the number of principally polarized abelian varieties with $a = g$. Such abelian varieties are sometimes called *superspecial*.

Corollary 9.5. *We have*

$$\sum_X \frac{1}{\#\mathrm{Aut}(X)} = (-1)^G 2^{-g}\Big[\prod_{j=1}^{g}(p^j+(-1)^j)\Big]\cdot\zeta(-1)\zeta(-3)\ldots\zeta(1-2g),$$

where the sum is over the isomorphism classes (over $\overline{\mathbb{F}}_p$) of principally polarized abelian varieties of dimension g with $a = g$.

Proof. Combine the formula for T_g with the Proportionality Theorem. $\square$

For each element w we now find a degeneracy locus $\mathcal{U}_w$. In particular we find such a locus $\mathcal{U}_\mu$ for each partition μ in $\mathcal{F}$ and the $\mathcal{U}_\mu$ actually lie in $\mathcal{U}_\emptyset$. The fact that T_g is zero-dimensional implies that the codimension of each $\mathcal{U}_\mu$ equals $\mathrm{codim}(w(\mu)) = |\mu|$. We can apply a theorem of Fulton to determine the class of the degeneracy locus U_μ in $CH_{\mathbb{Q}}^*(\mathcal{F})$.

For an admissible diagram $\mu = \{\mu_1 > \ldots > \mu_r > 0\}$ Fulton defines a determinant ('Schur function')

$$\Delta_\mu(x_i) = \det(x_{\mu_i+j-i})_{1\leq i,j\leq r};$$

this is a polynomial with integer coefficients in the commuting variables $x_1, x_2, \ldots$. Define a 'double Schubert function' by putting

$$\Delta(x,y) = \Delta_{g,g-1,\ldots,1}\big(\sigma_i(x_1,\ldots,x_g) + \sigma_i(y_1,\ldots,y_g)\big).$$

The operators ∂_i ('divided difference operators') on the ring $\mathbb{Z}[x_1, x_2, \ldots,]$ are defined by setting for $F(x) \in \mathbb{Z}[x_1, \ldots, x_g]$

$$\partial_i(F(x)) = \begin{cases} \frac{F(x)-F(s_i(x))}{x_i-x_{i+1}} & \text{if } i < g \\ \frac{F(x)-F(s_g(x))}{2x_g} & \text{if } i = g \end{cases}$$

Put

$$\Delta = \Delta(x,y) = \Delta_{g,g-1,\ldots,1}\big(\sigma_i(x_1,\ldots,x_g) + \sigma_i(y_1,\ldots,y_g)\big).$$

We shall apply the result of Fulton [7] and state an abstract formula for our cycle classes. In principle one then can calculate the pushforward algorithmically. But it seems difficult to get closed formulas for the cycle classes of these pushforwards.

We find:

Theorem 9.6. *Let μ be an admissible diagram whose corresponding element w_μ in the Weyl group is written as $s_{i_\ell} \cdots s_{i_1}$. The cycle class of the degeneracy locus U_μ in $CH^*_{\mathbb{Q}}(\mathcal{F})$ is given by*

$$u_\mu = \partial_{i_\ell} \cdots \partial_{i_1} \left(\left(\prod_{i+j \leq g} (x_i - y_i) \right) \cdot \Delta(x,y) \right) \Big|_{\{x_i = p^{l_i}, y_i = -l_i\}}$$

Corollary 9.7. *The pushforward of the class of $[\mathcal{U}_\mu]$ under π is given by $\pi_*(u_\mu)$ and is a multiple of the class of the reduced cycle Z_μ, the multiple being equal to the number of final filtrations refining the canonical filtration associated to μ. The class belongs to the tautological ring.*

We list here the formulas for $g = 3$. The multiplicity is given by the factor in square brackets.

Formulas for $g = 3$.

$$\pi_*(\{\emptyset\}) = [(1+p)(1+p+p^2)]$$

$$\pi_*(\{1\}) = [(1+p)] \times (1-p)\,\lambda_1$$

$$\pi_*(\{2\}) = (p-1)(p^2-1)\,\lambda_2$$

$$\pi_*(\{1,2\}) = [(1+p)] \times \{(1-p)(1+p^2)\,\lambda_1\,\lambda_2 - 2(-1+p^3)\,\lambda_3\}$$

$$\pi_*(\{3\}) = (p-1)(p^2-1)(p^3-1)\,\lambda_3$$

$$\pi_*(\{1,3\}) = [(1+p)] \times (-1+p)^2\,(1-p+p^2)\,\lambda_1\,\lambda_3$$

$$\pi_*(\{2,3\}) = (-1+p)^3\,(1+p)\,(-1+p-p^2)\,(1+p+p^2)\,\lambda_2\,\lambda_3$$

$$\pi_*(\{1,2,3\}) = [(1+p^3)] \times (p-1)\,(p^2+1)(p^3-1)\,\lambda_1\,\lambda_2\,\lambda_3.$$

The canonical filtration on $H^1_{dR}(X)$ is the most economical one with respect to the operation V. The types of this filtration correspond to the elements of $(\mathbb{Z}/2\mathbb{Z})^g$ in the Weyl group $(\mathbb{Z}/2\mathbb{Z})^g \rtimes S_g$. The other types of filtrations are obtained by applying an element of S_g to the right. Therefore, in the closure of a stratum $\mathcal{U}_\mu$ we find also $\mathcal{U}_w$ with $w \notin (\mathbb{Z}/2\mathbb{Z})^g$. This explains the phenomenon observed by Oort in [24], 4.6.

We can use Pieri-type formulas to get results on the boundary of strata. In particular one obtains the result of Ekedahl-Oort that the class λ_1 is torsion on the open strata. We refer to [3].

Some strata can be described in detail. Consider the locus associated to the partition

$$\mu = \{g, g-1, \ldots, 3, 2\}.$$

This is the penultimate stratum and it is of dimension 1. It is proved by Ekedahl and Oort that this locus is connected.

Define $h(g)$ by

$$h(1) = 1, \qquad h(g) = h(g-1) \times \frac{p^g + (-1)^g}{p + (-1)^g} \quad \text{for} \quad g > 1.$$

Then we have

$$\pi_*(U_{[g,g-1,\dots,1]}) = h(g)[T_g]$$

with $[T_g]$ given as above.

Theorem 9.8. *The cycle class of $\pi_*(\mathcal{U}_\mu)$ for $\mu = \{g,g-1,\dots,3,2\}$ is given by*

$$\frac{(p-1)}{(p^2+1)} \times h(g) \times \left(\prod_{i=1}^{g}(p^i + (-1)^i)\right) \times \lambda_2\lambda_3\dots\lambda_g.$$

In $\mathcal{U}_0$ this locus consists of a configuration of $\mathbb{P}^1$'s.

Since the degree of λ_1 on each copy of $\mathbb{P}^1$ is $p-1$ one can compute the number of components. This locus is highly reducible. But some loci are irreducible:

Theorem 9.9. *For all $a < g$ the locus T_a is irreducible. In particular, the locus $T_1 = V_{g-1}$ is irreducible.*

Proof. By Ekedahl-Oort (cf. [24]) we know that the locus corresponding to the diagram $\{g,g-1,\dots,3,2\}$ is connected. This implies that T_a is connected for $a < g$. We know that $\mathrm{Sing}(T_a) \subseteq T_{a+1}$, hence of codimension > 1. Actually, one can describe the normal bundle to $T_a - T_{a+1}$. By a theorem of Hartshorne (cf. [11]) this implies that $T_a - \mathrm{Sing}(T_a)$ is connected. $\square$

The special case of the irreducibility of V_{g-1} was also shown by Ekedahl and Oort, cf. [24], (6.5).

10   Effectivity of Tautological Classes

As a corollary of the results in the preceding section we obtain the effectivity of the expressions in the tautological ring that are the cycle classes of the Ekedahl-Oort strata. In particular we obtain the following result.

Theorem 10.1. *The Chern classes $\lambda_i \in CH_{\mathbb{Q}}(\mathcal{A}_g \otimes \mathbb{F}_p)$ of the Hodge bundle are represented by effective classes (with $\mathbb{Q}$-coefficients).*

Proof. The class of the locus of abelian varieties with p-rank $\leq f$ is a multiple of λ_{g-f}. $\square$

Remark 10.2. The Hodge bundle has an ample determinant bundle with first Chern class λ_1, but the bundle $\mathbb{E}$ is not positive in positive characteristic for $g \geq 2$. For example, for $g = 2$ the restriction of the Hodge bundle to a line from the p-rank 0 locus is isomorphic to $O(p) \oplus O(-1)$ which is evidently not positive. In general, $\mathbb{E}$ is not positive on the stratum defined by $\{g,g-1,\dots,3,2\}$.

11   Some Additional Results

We describe as an example the canonical type for hyperelliptic curves of 2-rank 0 in characteristic $p = 2$.

Lemma 11.1. *A hyperelliptic curve C of genus g and of 2-rank 0 over $k = \bar{k}$ of characteristic $p = 2$ can be written as*

$$y^2 + y = xP(x^2) = x(a_1x^2 + a_2x^4 + ... + a_{g-1}x^{2g-2} + x^{2g}).\qquad(13)$$

Theorem 11.2. *For a hyperelliptic curve as in (13) the canonical type is described by the corresponding partition $\mu = [g, g-2, g-4, ...]$. In particular, the canonical filtration is independent of the coefficients a_i.*

We refer for the proof to [3].

For $g = 1$ and $g = 2$ the supersingular locus coincides with V_0 and thus occurs as a stratum in the Ekedahl-Oort stratification. For $g \geq 3$ this is no longer true. The definition of supersingular locus involves also the higher filtrations, i.e. it uses not only $X[p]$, but the higher group schemes $X[p^i]$ as well. But in a number of cases we can determine the class of this locus. We state here the result for the case $g = 3$.

Theorem 11.3. *The class of the supersingular locus S_3 in $\mathcal{A}_3^* \otimes \mathbb{F}_p$ is*

$$[S_3] = (p-1)(p^2-1)(p^3-1)(p-1)(p^2+1)\lambda_1\lambda_3.$$

The proof uses the explicit description by Li and Oort of the moduli of principally polarized supersingular abelian varieties, cf. [15]. We refer the reader to [3] for the proof and formulas for other genera.

One can define analogues of the stratification defined here on $\mathcal{A}_g \otimes \mathbb{F}_p$ for other moduli spaces, e.g. for the moduli space of K3 surfaces of given genus in characteristic p. We refer to a joint paper with T. Katsura, see [9].

Bibliography

[1] A. Borel, J-P. Serre: Le théorème de Riemann-Roch (d'après Grothendieck). *Bull. Soc. Math. France* **86** (1958), 97–136.

[2] T. Ekedahl: On supersingular curves and abelian varieties. *Math. Scand.* **60** (1987), 151–178.

[3] T. Ekedahl, G. van der Geer: Manuscripts in preparation.

[4] T. Ekedahl, F. Oort : Connected subsets of a moduli space of abelian varieties. Preliminary version of a preprint.

[5] C. Faber, G. van der Geer: Complete subvarieties of $\mathcal{M}_g$ and the Prym map. In preparation.

[6] G. Faltings, C-L. Chai: Degeneration of abelian varieties. Ergebnisse der Math. 22. Springer Verlag 1990.

[7] W. Fulton: Determinantal formulas for orthogonal and symplectic degeneracy loci. *J. Diff. Geom.* **43** (1996), 276–290.

[8] G. van der Geer: The Chow ring of the moduli space of abelian threefolds. *J. of Alg. Geometry* **7** (1998), 753–770.

[9] G. van der Geer, T. Katsura: A stratification of the moduli space of K3 surfaces in positive characteristic. In preparation.

[10] G. van der Geer, F. Oort: Moduli of abelian varieties : a short introduction and survey. This volume, 1–21.

[11] R. Hartshorne: Complete intersections and connectedness. *Amer. J. Math.* **84** (1962), 497–508.

[12] F. Hirzebruch: Automorphe Formen und der Satz von Riemann-Roch. In *Symposium Internacional de Topologia Algebrica* (México 1956), 129–144. México: La Universidad Nacional Autónoma de México 1958. (= Collected Papers I , Springer Verlag, p. 345)

[13] F. Hirzebruch: Kommentar zu 'Elliptische Differentialoperatoren auf Mannigfaltigkeiten'. In: Collected Papers II, Springer Verlag, p. 773.

[14] N. Koblitz: *p*-adic variation of the zeta function over families of varieties defined over finite fields. *Compositio Math.* **31**, (1975), 119–218.

[15] K-Z. Li, F. Oort: Moduli of supersingular abelian varieties. Lecture Notes in Mathematics 1680, Springer Verlag, Berlin-New York, 1998.

[16] L. Moret-Bailly: Pinceaux de variétés abéliennes. Astérisque 129 (1985).

[17] D. Mumford: Picard groups of moduli problems. In *Arithmetical Algebraic Geometry*. Ed. O.F.G. Schilling. Harper and Row, 1965, p. 33–81.

[18] D. Mumford: Hirzebruch's Proportionality Theorem in the non-compact case. *Inv. Math.* **42** (1977) , 239–272.

[19] D. Mumford: On the Kodaira dimension of the Siegel modular variety. In: Algebraic Geometry—Open Problems. SLNM 997, 348–375.

[20] P. Norman, F. Oort: Moduli of abelian varieties. *Ann. Math.* **112** (1980), 413–439.

[21] T. Oda : The first de Rham cohomology group and Dieudonné modules. Ann. Sci. ENS (1969), 63–135.

[22] F. Oort: Subvarieties of moduli spaces. *Inv. Math.* **24** (1974), 95–119.

[23] F. Oort: Complete subvarieties of moduli spaces. In: *Abelian Varieties* (W. Barth, K. Hulek, H. Lange, eds.), de Gruyter Verlag, Berlin, 1995, p. 225–235.

[24] F. Oort: A stratification of a moduli space of polarized abelian varieties in positive characteristic. Preprint May 1996. This volume, 47–64.

[25] P. Pragacz: Algebro-geometric applications of Schur S- and Q- polynomials. In: Topics in Invariant Theory. Séminaire d'Algèbre Dubreil-Malliavin 1989–1990. SLNM 1478, 130–191 (1991).

[26] P. Pragacz, J. Ratajski: Formulas for Lagrangian and orthogonal degeneracy loci; the $\tilde{Q}$-polynomial approach. *Comp. Math.* **107** (1997), 11–87.

Locally Symmetric Families of Curves and Jacobians

Richard Hain

Abstract

The moduli space $\mathcal{A}_g$ of principally polarized abelian varieties of dimension g is a locally symmetric variety. Denote the closure in $\mathcal{A}_g$ of the locus of jacobians by $\mathcal{J}_g$. In this paper we make a preliminary investigation of locally symmetric subvarieties X of $\mathcal{A}_g$ that are contained in $\mathcal{J}_g$ and contain the moduli point of the jacobian of a smooth curve. Under certain hypotheses (X is "simple", the corresponding family of abelian varieties can be lifted to a family of curves and a rank condition), we prove that such an X has to be a ball quotient. Our main tools are group cohomology and naive geometric considerations.

1 Introduction

In this paper we study locally symmetric families of curves and jacobians. For the purposes of this paper, a jacobian is an abelian variety that is Pic^0 of a semi-stable curve. Denote the moduli space of principally polarized abelian varieties of dimension g by $\mathcal{A}_g$. It is a locally symmetric variety. By a *locally symmetric family of jacobians*, we mean a family of jacobians parameterized by a locally symmetric variety X where the period map $X \to \mathcal{A}_g$ is a map of locally symmetric varieties — cf. (8.2). A *locally symmetric family of curves* is a family of semi-stable curves over a locally symmetric variety where Pic^0 of each curve in the family is an abelian variety and where the corresponding family of jacobians is a locally symmetric family — cf. (7.1). Not every locally symmetric family of jacobians can be lifted to a locally symmetric family of curves, even if one passes to arbitrary finite unramified covers of the base, as can be seen by looking at the universal abelian variety of dimension 3 — see (8.3).

Definition 1.1. A locally symmetric variety X is *bad* if it has a locally symmetric divisor. Otherwise, we shall say that X is *good*.

Each locally symmetric variety X arises from a semisimple $\mathbb{Q}$-group G whose associated symmetric space is hermitian. We may suppose that G is simply connected as a linear algebraic group. In this case we can write G as a product $G_1 \times \cdots \times G_m$ of almost simple $\mathbb{Q}$-groups. There is a corresponding splitting $X' = X_1 \times \cdots \times X_m$ of a finite covering X' of X, and X is good if and only if each X_i is good. Since each G_i is almost simple, each X_i is also "simple" in the sense that it has no finite cover which splits as a product of locally symmetric varieties.

Remark 1.2.   (i) Suppose that X is a locally symmetric variety whose associated $\mathbb{Q}$-group G is almost simple. Nolan Wallach [24] has communicated to me a proof that if the $\mathbb{Q}$-rank of G is ≥ 3, then X is good.

(ii) It follows from the classification of hermitian symmetric spaces [12]p. 518 that if the non-compact factor of $G(\mathbb{R})$ is simple and not $\mathbb{R}$-isogenous to $SO(n,2)$ or $SU(n,1)$ for any n, then X is also good — see (5.2).

Recall that the symmetric space $SU(n,1)/U(n)$ is the complex n-ball.

Theorem 1. *Suppose that $p : C \to X$ is a non-constant locally symmetric family of curves over the locally symmetric variety X. Suppose that the corresponding $\mathbb{Q}$-group is almost simple. If every fiber of p is smooth, or if X is good and the generic fiber of p is smooth, then X is a quotient of the complex n-ball.*

Note that the simplicity of the group G implies that the period map $X \to A_g$ is finite.

After studying the geometry of the locus of jacobians, we are able to prove the following result about a locally symmetric family $B \to X$ of jacobians. In this result, X^{dec} denotes the set of points of X where the fiber is the jacobian of a singular curve and X^* denotes its complement. The locus of hyperelliptic jacobians in X will be denoted by $X_{\mathcal{H}}$, and $X^* \cap X_{\mathcal{H}}$ will be denoted by $X_{\mathcal{H}}^*$.

Theorem 2. *Suppose that $B \to X$ is a non-constant locally symmetric family of jacobians and that X^* is non-empty. Suppose that the corresponding $\mathbb{Q}$-group is almost simple. If X is good, then either:*

(i) X is a quotient of the complex n-ball, or

(ii) $g \geq 3$, each component of X^{dec} is of complex codimension ≥ 2, $X_{\mathcal{H}}^*$ is a non-empty divisor in X^* which is smooth if $g > 3$, and the family does not lift to a locally symmetric family of curves.

Of course, we can also apply this theorem to all locally symmetric subvarieties of X. Note that the second case in the theorem does occur; take $g = 3$ and X to be $\mathcal{A}_3$ (cf. (8.3)). I do not know any examples where the second case of the theorem holds and $g > 3$, but suspect there are none.

These results depend upon the following kind of rigidity result for mapping class groups. We shall denote the mapping class group associated to a surface of genus g with n marked points and r non-zero boundary components by $\Gamma_{g,r}^n$.

Theorem 3. *Suppose that G is a simply connected, almost simple $\mathbb{Q}$-group, that the symmetric space associated to $G(\mathbb{R})$ is hermitian, and that Γ is an arithmetic subgroup of G. If $G(\mathbb{R})$ has real rank > 1 (that is, it is not $\mathbb{R}$-isogenous to the product of $SU(n,1)$ with a compact group) and if $g \geq 3$, then the image of every homomorphism $\Gamma \to \Gamma_{g,r}^n$ is finite.*

This is a special case of a much more general result which was conjectured by Ivanov and proved by Farb and Masur [6]. (They do not assume hermitian symmetric nor that Γ is arithmetic.) Their methods are very different from ours, so we have included a proof of this theorem in the case when $r + n > 0$, and a weaker statement that is sufficient for our applications in the case $r + n = 0$.

Our approach is via group cohomology. The basic idea is that if $C \to X$ is a locally symmetric family of curves, then there is a homomorphism from $\pi_1(X, *)$ to the mapping class group Γ_g, which is the orbifold fundamental group of the moduli space of curves. The rigidity theorem above shows that there are very few homomorphisms from arithmetic groups to mapping class groups. To get similar results for locally symmetric families of jacobians, we exploit the fact that the fundamental group of a smooth variety does not change when a subvariety of codimension ≥ 2 is removed. It is for this reason that we are interested in good locally symmetric varieties.

This work arose out of a question of Frans Oort, who asked if there are any positive dimensional Shimura subvarieties of the jacobian locus not contained in the locus of reducible jacobians. He suspects that once the genus is sufficiently large, there may be none. His interest stems from a conjecture of Coleman [5]Conj. 6 which asserts that the set of points in $\mathcal{M}_g$ whose jacobians have complex multiplication is finite provided that $g \geq 4$. This is false as stated as was shown by de Jong and Noot [15] who exhibited infinitely many smooth curves of genus 4 and genus 6 whose jacobians have complex multiplication. (Note that when $g \leq 3$ there are infinitely many curves whose jacobian has complex multiplication as $\mathcal{M}_g$ is dense in $\mathcal{A}_g$ in these cases.) However, Oort tentatively believes Coleman's conjecture may be true when g is sufficiently large. Partial results on Oort's question, which are complementary to results in this paper, have been obtained by Ciliberto, van der Geer and Teixidor i Bigas in [3] and [4].

I am very grateful to Frans Oort for asking me this question, and also to my colleagues Les Saper, Mark Stern (especially) and Jun Yang for helpful discussions about symmetric spaces, arithmetic groups and their cohomology. I would also like to thank Nolan Wallach for communicating his result to me, and Nicholai Ivanov for telling me about his belief that rigidity for mapping class groups held in this form. Finally, thanks to Ben Moonen, Carel Faber and Gerard van der Geer for their very helpful comments on this manuscript, which helped improve the exposition and saved me from many a careless slip.

2   Background and Definitions

Here we gather together some definitions to help the reader. A *locally symmetric variety* is a locally symmetric space of the form

$$\Gamma\backslash G(\mathbb{R})/K$$

where G is a semi-simple $\mathbb{Q}$-group, Γ is an arithmetic subgroup of G, K is a maximal compact subgroup of $G(\mathbb{R})$, and the symmetric space $G(\mathbb{R})/K$ is hermitian. The moduli space of abelian varieties with a level l structure

$$\mathcal{A}_g[l] = Sp_g(\mathbb{Z})[l]\backslash Sp_g(\mathbb{R})/U(g)$$

is a primary example. Here $Sp_g(\mathbb{Z})[l]$ denotes the level l subgroup of $Sp_g(\mathbb{Z})$. This moduli space is well known to be a quasi-projective variety. A fundamental theorem of Baily and Borel [1] asserts that every locally symmetric variety is a quasi-projective algebraic variety. The imbedding into projective space is given by automorphic forms.

Suppose that

$$X_1 = \Gamma_1\backslash G_1(\mathbb{R})/K_1 \text{ and } X_2 = \Gamma_2\backslash G_2(\mathbb{R})/K_2$$

are two locally symmetric varieties. A map of locally symmetric varieties $X_1 \to X_2$ is a map induced by a homomorphism of $\mathbb{Q}$-algebraic groups $G_1 \to G_2$.

Recall that a semi-simple k-group G is *almost simple* if its adjoint form is simple. An algebraic k-group is *absolutely (almost) simple* if the corresponding group over $\overline{k}$ is (almost) simple.

Every simply connected $\mathbb{Q}$-group G is the product of simply connected, almost simple $\mathbb{Q}$-groups:

$$G = \prod_{i=0}^{n} G_i$$

This implies that if Γ is an arithmetic subgroup of G, then

$$\prod_{i=0}^{n} \Gamma \cap G_i$$

has finite index in Γ. It follows that if X is the locally symmetric space associated to Γ and X_i the locally symmetric space associated to Γ_i, then the natural mapping

$$\prod_{i=0}^{n} X_i \to X$$

is finite. If X is a locally symmetric variety (i.e., $G(\mathbb{R})/K$ is hermitian), then each X_i will also be a locally symmetric variety.

We shall call a locally symmetric variety associated to an arithmetic subgroup of an almost simple $\mathbb{Q}$-group a *simple* locally symmetric variety. They have no finite covers which are products of locally symmetric varieties. Also, since the associated $\mathbb{Q}$-group is almost simple, every map from a simple locally symmetric variety to a locally symmetric variety will be either constant or finite.

We shall make critical use of a vanishing theorem proved by Raghunathan [21]. Suppose that Γ is an arithmetic subgroup of an almost simple $\mathbb{Q}$-group G. If $G(\mathbb{R})$ has real rank ≥ 2, then

$$H^1(\Gamma, V) = 0$$

for all finite dimensional representations of $G(\mathbb{R})$.

The classification of hermitian symmetric spaces [12]p. 518 implies that if

$$X = \Gamma \backslash G(\mathbb{R})/K$$

is a simple locally symmetric variety where $G(\mathbb{R})$ has real rank 1, then the associated symmetric space $G(\mathbb{R})/K$ is the complex n-ball $SU(n,1)/U(n)$. This rank condition, and the corresponding failure of Raghunathan's Vanishing Theorem, is the principal reason we cannot handle ball quotients.

Margulis's Rigidity Theorem [17] has the following important consequence for mappings between locally symmetric varieties. Suppose that X_1 and X_2 are locally symmetric varieties as above and that X_1 is simple. If the real rank of $G_1(\mathbb{R})$ is 2 or more, then for every homomorphism $\phi : \Gamma_1 \to \Gamma_2$, there is a finite covering

$$\begin{array}{ccc} \Gamma_1' \backslash G_1(\mathbb{R})/K_1 & \longrightarrow & \Gamma_1 \backslash G_1(\mathbb{R})/K_1 \\ \| & & \| \\ X_1' & \longrightarrow & X_1 \end{array}$$

of locally symmetric varieties corresponding to a finite index subgroup Γ_1' of Γ_1, and a map of locally symmetric varieties $X_1' \to X_2$ which induces the restriction

$$\phi|_{\Gamma_1'} : \Gamma_1' \to \Gamma_2$$

of ϕ to Γ_1'.

Every arithmetic group Γ has a torsion free subgroup of finite index. When Γ is torsion free, the locally symmetric variety $\Gamma \backslash G(\mathbb{R})/K$ is a $K(\Gamma,1)$. That is, it has fundamental group Γ and all of its higher homotopy groups are trivial.

The level l subgroup $\Gamma_{g,r}^n[l]$ of the mapping class group $\Gamma_{g,r}^n$ is the kernel of the natural homomorphism $\Gamma_{g,r}^n \to Sp_g(\mathbb{Z}/l\mathbb{Z})$. It is torsion free when $l \geq 3$. In this case it is isomorphic to the fundamental group of $\mathcal{M}_{g,r}^n[l]$, the moduli space of smooth projective curves of genus g over $\mathbb{C}$ with n marked points, r non-zero tangent vectors and a level l structure. (As is customary, we shall omit r and n when they are zero.) This isomorphism is unique up to an inner automorphism. We shall often fix a level $l \geq 3$ to guarantee that $\mathcal{A}_g[l]$ and $\mathcal{M}_{g,r}^n[l]$ are smooth.

3   Homomorphisms to Mapping Class Groups

Suppose that Γ is a discrete group and that $\phi : \Gamma \to \Gamma^n_{g,r}$ is a homomorphism. Denote the composite of ϕ with the natural homomorphism $\Gamma^n_{g,r} \to Sp_g(\mathbb{Z})$ by ψ. Denote the kth fundamental representation of $Sp_g(\mathbb{R})$ by V_k $(1 \leq k \leq g)$. Each of these can be viewed as a Γ module via ψ.

Theorem 3.1. *If $H^1(\Gamma,V_3) = 0$ and $g \geq 3$, then $\psi^* : H^2(Sp_g(\mathbb{Z}),\mathbb{Q}) \to H^2(\Gamma,\mathbb{Q})$ vanishes.*

Proof. It suffices to prove the result when $r = n = 0$. I'll give a brief sketch. We can consider the group $H^1(\Gamma_g,V_3)$. One has the cup product map

$$H^1(\Gamma_g,V_3)^{\otimes 2} \to H^2(\Gamma_g,\Lambda^2 V_3).$$

Since V_3 has an Sp_g invariant, skew symmetric bilinear form, we have a map

$$H^2(\Gamma_g,\Lambda^2 V_3) \to H^2(\Gamma_g,\mathbb{R}).$$

We therefore have a map

$$c : H^1(\Gamma_g,V_3)^{\otimes 2} \to H^2(\Gamma_g,\mathbb{R}).$$

It follows from Dennis Johnson's results [14] (cf. [9](5.2)) that the left hand group has dimension 1 when $g \geq 3$, and from [8]§§7–8 that this map is injective. It is also known that the natural map

$$H^2(Sp_g(\mathbb{Z}),\mathbb{R}) \to H^2(\Gamma_g,\mathbb{R})$$

is an isomorphism for all $g \geq 3$ and that both groups are isomorphic to $\mathbb{R}$ (cf. [11], [18], [19].) The result now follows from the commutativity of the diagram.

$$
\begin{array}{ccc}
H^1(\Gamma_g,V_3)^{\otimes 2} & \xrightarrow{\ c\ } & H^2(\Gamma_g,\mathbb{R}) \\
\downarrow & & \downarrow \\
H^1(\Gamma,V_3)^{\otimes 2} & \xrightarrow{\ c\ } & H^2(\Gamma,\mathbb{R})
\end{array}
$$

$\square$

Corollary 3.2. *For all $g \geq 3$ and all $l > 0$, the homomorphism $\Gamma_g[l] \to Sp_g(\mathbb{Z})[l]$ does not split.*

Proof. By a result of Raghunathan [21], if $g \geq 2$, Γ is a finite index subgroup of $Sp_g(\mathbb{Z})$ and V is a rational representation of $Sp_g(\mathbb{R})$, then $H^1(\Gamma,V)$ vanishes. In particular,

$$H^1(Sp_g(\mathbb{Z})[l],V_3) = 0$$

when $g \geq 3$. If there were a splitting of the canonical homomorphism $\Gamma_g[l] \to$ $Sp_g(\mathbb{Z})[l]$, then, by Theorem 3.1, the homomorphism

$$H^2(Sp_g(\mathbb{Z}),\mathbb{Q}) \to H^2(Sp_g(\mathbb{Z})[l],\mathbb{Q})$$

induced by the inclusion would be trivial. But when $g \geq 3$, Borel's stability theorem [2] (see also [10]3.1) implies that this mapping is an isomorphism. It follows that no such splitting can exist.   □

Remark 3.3. There are cases when the cohomological obstruction above vanishes but there is still no lift. An example is given in the appendix.

Suppose that X is a topological space with fundamental group π and that $\mathbb{V}$ is the local system corresponding to the π module V. It is a standard fact that there is a natural map

$$H^k(\pi,V) \to H^k(X,\mathbb{V})$$

which is an isomorphism when $k \leq 1$ and injective when $k = 2$. We thus have the following corollary.

Corollary 3.4. *Suppose that $g \geq 3$, $l \geq 3$ and that $\phi : X \to \mathcal{M}_g[l]$ is a continuous map from a topological space X to $\mathcal{M}_g[l]$. If $H^1(X,\mathbb{V}_3) = 0$, then the map*

$$H^2(\mathcal{A}_g,\mathbb{Q}) \to H^2(X,\mathbb{Q})$$

induced by the composition $X \to \mathcal{A}_g$ of ϕ with the period map vanishes.   □

4   Maps of Lattices to Mapping Class Groups

In this section we apply the results of the previous section to the case where Γ is the arithmetic group associated to a locally symmetric variety. We begin by recalling a result of Yang [25].

Suppose that G is a semisimple group over a number field k. Yang [25](3.3) defined an invariant $\ell(G/k)$ of G as follows. Fix a minimal parabolic P of G. Let T be the corresponding maximal k-split torus. Denote the relative root system of G by Φ, the base of Φ corresponding to P by Δ, and the set of positive relative roots by Φ_+. Set

$$\rho_P = \frac{1}{2} \sum_{\alpha \in \Phi_+} \alpha.$$

Then we can define

$$\ell(G/k) = \max \Big\{ q \in \mathbb{N} : 2\rho_P - \sum_{\alpha \in J} \alpha \text{ is strictly dominant}$$

$$\text{for all subsets } J \text{ of } \Phi_+ \text{ of cardinality} < q \Big\}.$$

Denote the Lie algebra of G by $\mathfrak{g}$ and its maximal compact subgroup by K. The importance of this invariant lies in the following result of Yang [25].

Theorem 4.1 (Yang). *If G is a semisimple $\mathbb{Q}$-group and Γ an arithmetic subgroup of G, then the natural map $H^m(\mathfrak{g},K) \to H^m(\Gamma,\mathbb{R})$ is injective whenever $m \leq \ell(G/\mathbb{Q})$.*   □

We will need the following consequence:

Lemma 4.2. *Suppose that Γ is an arithmetic subgroup of an almost simple $\mathbb{Q}$-group G. If $G \neq SL_2(\mathbb{Q})$, then $H^2(\mathfrak{g},K) \to H^2(\Gamma,\mathbb{R})$ is injective.*

Proof. When the $\mathbb{Q}$-rank of G is zero, the corresponding locally symmetric space is compact and the relative Lie algebra cohomology injects. So we suppose that G has positive $\mathbb{Q}$-rank. Note that $\ell(G/\mathbb{Q}) \geq 1$ for all such G.

Every almost simple $\mathbb{Q}$-group is of the form $R_{k/\mathbb{Q}}G$, where G is an absolutely almost simple k-group ([23]p. 46). It is not difficult to see that if G is a k-group, then

$$\ell((R_{k/\mathbb{Q}}G)/\mathbb{Q}) \geq [k : \mathbb{Q}]\,\ell(G/k).$$

So, by Yang's Theorem, it suffices to prove that for every absolutely almost simple group G, $\ell(G/k) \geq 2$. This follows from Tits' classification of the decorated Dynkin diagrams of such groups [23]pp 54–61 as the Dynkin diagram of each of these reduced root systems is connected and has ≥ 2 vertices — use [13]§10.2.   □

Proof of Theorem 3. [1] Since every simply connected almost simple $\mathbb{Q}$-group G is of the form $R_{k/\mathbb{Q}}G'$ where k is a number field and G' is a simply connected, absolutely almost simple k-group, the classification of Hermitian symmetric spaces [12]p. 518 implies that either $G(\mathbb{R})$ is isogenous to the product of $SU(n,1)$ and a compact group, or has real rank ≥ 2.

By the superrigidity theorem of Margulis [17], we know that, after replacing Γ by a finite index subgroup if necessary, there is a $\mathbb{Q}$-group homomorphism $G \to Sp_g(\mathbb{Q})$ that induces the homomorphism $\Gamma \to \Gamma^n_{g,r} \to Sp_g(\mathbb{Z})$. It follows that the representation V_3 of Γ is the restriction of a representation of G. Since the real rank of G is at least 2, it follows from Raghunathan's result [21] that $H^1(\Gamma,V_3)$ also vanishes. So, by (3.1), $H^2(Sp_g(\mathbb{Z}),\mathbb{R}) \to H^2(\Gamma,\mathbb{R})$ vanishes. But we have the commutative diagram

$$
\begin{array}{ccc}
H^2(Sp_g(\mathbb{Z}),\mathbb{C}) & \longrightarrow & H^2(\Gamma,\mathbb{C}) \\[4pt]
\| & & \uparrow{\scriptstyle p} \\[4pt]
H^2(\mathfrak{sp}_g(\mathbb{R}),U(g);\mathbb{C}) & \longrightarrow & H^2(\mathfrak{g}(\mathbb{R}),K;\mathbb{C}) \\[4pt]
\| & & \| \\[4pt]
H^2(X,\mathbb{C}) & \longrightarrow & H^2(Y,\mathbb{C})
\end{array}
$$

---

[1] In the case $r + n = 0$ we shall prove that the image of $\Gamma \to \Gamma_g \to Sp_g(\mathbb{Q})$ is finite. If $\Gamma \to \Gamma_g$ is the monodromy representation of a family of jacobians, the finiteness of the image of this homomorphism implies the isotriviality of the family. This is all we shall need in the sequel.

where X is the compact dual of Siegel space $Sp_g(\mathbb{R})/U(g)$, Y is the compact dual of $G(\mathbb{R})/K$, and the vertical maps are the standard ones. Consider the homomorphism $G \to Sp_g(\mathbb{R})$. If it is trivial, then the image of Γ in $Sp_g(\mathbb{Z})$ is finite. (Remember that we passed to a finite index subgroup earlier in the argument.) We will show that it must be trivial.

Since $G(\mathbb{R})$ is semisimple, it has a finite cover which is a product of simple groups G_i. Since the symmetric space associated to $G(\mathbb{R})$ is hermitian, the symmetric space associated to each G_i is also hermitian. Denote the compact dual of the symmetric space associated to G_i by Y_i. Then Y is the product of the Y_i. Since each G_i is simple, the induced map $G_i \to Sp_g(\mathbb{R})$ is either trivial or has finite kernel. It follows that the corresponding maps $Y_i \to X$ of compact duals are either injective or trivial. If $G \to Sp_g(\mathbb{R})$ is non-trivial, there is an i such that $G_i \to Sp_g(\mathbb{R})$ is injective. The corresponding map $Y_i \to X$ of compact duals is injective. Since X and Y_i are Kähler, and since $H^2(Sp_g(\mathbb{Z}),\mathbb{C})$ is one dimensional, the map $H^2(X,\mathbb{C}) \to H^2(Y_i,\mathbb{C})$ must be injective as the Kähler form cannot vanish on Y_i. It follows that the bottom map in the diagram is also injective. Combining this with (4.2), we see that the map p is injective, which in turn implies that the top map is injective. But this contradicts (3.1). So $G \to Sp_g(\mathbb{R})$ must be trivial. This is the statement we set out to prove when $r + n = 0$.

Since $G \to Sp_g(\mathbb{R})$ is trivial, we conclude that Γ maps to the Torelli group $T_{g,r}^n$. But this group is residually torsion free nilpotent (i.e., injects into its unipotent completion — see [10](14.9)) when $r + n > 0$. Since $H_1(\Gamma,\mathbb{Q}) = 0$, the image of Γ in every unipotent group over $\mathbb{Q}$ is trivial. It follows that the image of Γ in the unipotent completion of $T_{g,r}^n$, and therefore in $T_{g,r}^n$, is trivial.[2] $\square$

Corollary 4.3. *If we have a non-constant family of non-singular curves of genus ≥ 3 over a locally symmetric variety whose corresponding $\mathbb{Q}$-group G is almost simple, then the base is a quotient of the complex n-ball.*

Proof. The base X of the family is $X = \Gamma\backslash G(\mathbb{R})/K$ where G is the associated $\mathbb{Q}$-group, Γ is an arithmetic subgroup of G, and K is a maximal compact subgroup of $G(\mathbb{R})$. By passing to a finite index subgroup if necessary, we may assume that Γ is torsion free. Since the family is non trivial, the homomorphism $\Gamma \to Sp_g(\mathbb{Z})$ induced by the period map of the family is not finite. It follows from Theorem 3.1 (see also the footnote to its proof) that $G(\mathbb{R})$ must have real rank 1. This implies that $G(\mathbb{R})/K$ is the complex n-ball. $\square$

---

[2] It would be interesting to know if T_g is residually torsion free nilpotent when g is sufficiently large.

5 Locally Symmetric Hypersurfaces in Locally Symmetric Varieties

In order to extend (4.3) to locally symmetric families of stable curves, we will need to know when a locally symmetric variety has a locally symmetric hypersurface. Recall that every almost simple $\mathbb{Q}$-group is of the form $R_{k/\mathbb{Q}}G$ where k is a number field and G is an absolutely almost simple group over k.

Proposition 5.1. *If $G' = R_{k/\mathbb{Q}}G$, where G is a k-group, then $G(\mathbb{R})$ is a hermitian symmetric space if and only if k is a totally real field and the symmetric space associated to each real imbedding of k into $\mathbb{R}$ is hermitian.*

Proof. We have

$$G'(\mathbb{R}) = \prod_{\nu:k\to\mathbb{C}} G_\nu.$$

where ν ranges over complex conjugate pairs of imbeddings of k into $\mathbb{C}$. Note that G_ν is an absolutely simple real group if ν is a real imbedding and G_ν is not an absolutely simple real group if ν is not real. The symmetric space associated to $G'(\mathbb{R})$ is the product of the symmetric spaces of the G_ν. It is hermitian symmetric if and only if each of its factors is. But if ν is not real, then G_ν cannot be compact, and its symmetric space is not hermitian (see the list in Helgason [12]p. 518). So k is totally real.   □

Proposition 5.2. *Suppose that G is a simple real Lie group whose associated symmetric space is hermitian. If G is not isogenous to $SO(n,2)$ or $SU(n,1)$ for any n, then the symmetric space associated to G has no hermitian symmetric hypersurfaces.*

Proof. This follows from the classification of irreducible hermitian symmetric spaces [12]p. 518 using [22]Theorem 10.5.   □

6 Geometry of the Jacobian Locus

The geometry of the locus of jacobians will play a significant role in the sequel. The *jacobian locus* $\mathcal{J}_g[l]$ in $\mathcal{A}_g[l]$ is the closure of the image of the period map $\mathcal{M}_g[l] \to \mathcal{A}_g[l]$. Denote the locus of jacobians of singular curves by $\mathcal{J}_g^{\mathrm{dec}}[l]$, and the locus $\mathcal{J}_g[l] - \mathcal{J}_g^{\mathrm{dec}}[l]$ of jacobians of smooth curves by $\mathcal{J}_g^*[l]$. In this section we are concerned with the geometry of $\mathcal{J}_g[l]$ along the hyperelliptic locus of $\mathcal{J}_g^*[l]$ and along $\mathcal{J}_g^{\mathrm{dec}}[l]$.

Proposition 6.1. *If $g \geq 3$ and $l \geq 3$, then the locus of hyperelliptic jacobians in $\mathcal{J}_g^*[l]$ is smooth, and its projective normal cone in $\mathcal{A}_g[l]$ at the point $[C]$ is $\mathbb{P}(S^2(V_C))$, where V_C is a vector space of dimension $g - 2$. Further, its projective normal cone in $\mathcal{J}_g^*[l]$ at the point $[C]$ is $\mathbb{P}(V_C)$, which is imbedded in $\mathbb{P}(S^2V_C)$ via the "Plücker imbedding."*

Proof. This follows directly from [20].   □

The following fact must be well known.

Lemma 6.2. *If V is a complex vector space, then there is no line L in $\mathbb{P}(S^2V)$ that is contained in the image of the Veronese imbedding*

$$\nu : \mathbb{P}(V) \hookrightarrow \mathbb{P}(S^2V).$$

Proof. Denote the hyperplane class of $\mathbb{P}(V)$ by H_1 and that of $\mathbb{P}(S^2V)$ by H_2. Since a hyperplane section of $\mathbb{P}(V)$ in $\mathbb{P}(S^2V)$ is a quadric, $\nu^*H_2 = 2H_1$. If L is a line in $\mathbb{P}(S^2V)$ that is contained in $\mathbb{P}(V)$, then

$$H_2|_L = 2H_1|_L \in H^2(L,\mathbb{Z}).$$

This is impossible as $H_2|_L$ generates $H^2(L,\mathbb{Z})$. So no such line can exist.   □

Combined, these two results give us geometric information about how smooth subvarieties of $\mathcal{A}_g[l]$ that are contained in $\mathcal{J}_g[l]$ intersect the hyperelliptic locus.

Proposition 6.3. *Suppose that $g \geq 4$ and $l \geq 3$ and that Z is a connected complex submanifold of $\mathcal{A}_g[l]$. If Z is contained in $\mathcal{J}_g^*[l]$, then one of the following holds:*

(i) Z is contained in the hyperelliptic locus,

(ii) Z does not intersect the hyperelliptic locus,

(iii) the intersection of Z with the hyperelliptic locus is a smooth subvariety of Z of pure codimension 1.

Proof. Suppose that Z is neither contained in the locus of hyperelliptic jacobians nor disjoint from it. Suppose that P is in the intersection of Z and the hyperelliptic locus. Denote by V the $g - 2$ dimensional vector space corresponding to the point P as in (6.1). Denote the quotient of the tangent space T_PZ by the Zariski tangent space of the intersection of Z with the hyperelliptic jacobians by N_P. Since Z is contained in $\mathcal{J}_g^*[l]$, it follows from (6.1) that

$$\mathbb{P}(N_P) \hookrightarrow \mathbb{P}(V) \overset{\nu}{\hookrightarrow} \mathbb{P}(S^2V).$$

Since Z is a smooth subvariety of $\mathcal{A}_g[l]$, $\mathbb{P}(N_P)$ is a linear subspace of $\mathbb{P}(S^2V)$. It follows from (6.2) that N_P has dimension 0 or 1. If the dimension of N_P is 0 for generic P, then Z is contained in the hyperelliptic locus. Since Z is not contained in the hyperelliptic locus, it must be that N_P is one dimensional for generic P in the intersection. But since the dimension of N_P is bounded by 1, it is one for all P in the intersection of Z with the locus of hyperelliptic jacobians. This implies that the intersection is a smooth divisor.   □

Denote the point in $\mathcal{A}_h[l]$ corresponding to the principally polarized abelian variety A with level l structure by $[A]$. If $g' + g'' = g$, there is a generically finite-to-one morphism[3] $\mu : \mathcal{A}_{g'}[l] \times \mathcal{A}_{g''}[l] \to \mathcal{A}_g[l]$. It takes $([A'],[A''])$ to $[A' \times A'']$. Suppose that $[A' \times A'']$ is a simple point of the image of μ. Let

$$V' = H^{0,1}(A') \text{ and } V'' = H^{0,1}(A'')$$

Lemma 6.4. *The projective normal cone to the image of* $\operatorname{im}\mu$ *at* $[A \times B]$ *is naturally isomorphic to* $\mathbb{P}(V' \otimes V'')$.

Proof. This follows directly by looking at period matrices. $\square$

Now suppose that C is a genus g stable curve with one node and whose normalization is $C' \cup C''$ where $g(C') = g'$ and $g(C'') = g''$. Provided the level l is sufficiently large, $[\operatorname{Jac} C]$ will be a simple point of $\operatorname{im}\mu$. We now continue the discussion above with $A' = \operatorname{Jac} C'$ and $A'' = \operatorname{Jac} C''$. Note that we have canonical imbeddings

$$C' \to \mathbb{P}(V') \text{ and } C'' \to \mathbb{P}(V'').$$

Composing the product of these with the Segre imbedding

$$\mathbb{P}(V') \times \mathbb{P}(V'') \to \mathbb{P}(V' \otimes V'')$$

we obtain a morphism

$$\delta : C' \times C'' \to \mathbb{P}(V' \otimes V'') \tag{6.1}$$

which is a finite onto its image when both g' and g'' are 2 or more, and is a projection onto the second factor when $g' = 1$.

We can project the tangent cone of $\mathcal{J}_g[l]$ into the normal bundle of the image of $\mathcal{A}_{g'}[l] \times \mathcal{A}_{g''}[l]$ in $\mathcal{A}_g[l]$. We shall call this image the *relative normal cone* of $\mathcal{J}_g[l]$ in $\mathcal{A}_g[l]$. We can then projectivize to get a map from the projectivized relative normal cone of $\mathcal{J}_g[l]$ to the projective normal cone of $\mathcal{A}_{g'}[l] \times \mathcal{A}_{g''}[l]$ in $\mathcal{A}_g[l]$.

Proposition 6.5. *The image of the projectivized relative normal cone of* $\mathcal{J}_g[l]$ *in* $\mathcal{A}_g[l]$ *at* $\operatorname{Jac}(C' \times C'')$ *in the projective normal cone of* $\mathcal{A}_{g'}[l] \times \mathcal{A}_{g''}[l]$ *in* $\mathcal{A}_g[l]$ *at* $\operatorname{Jac}(C' \times C'')$ *is the image of the map (6.1).*

Proof. This follows directly from [7]Cor. 3.2. $\square$

Suppose that Z is a complex submanifold of $\mathcal{A}_g[l]$ which is contained in the jacobian locus. Set $Z^{\mathrm{dec}} = Z \cap \mathcal{J}^{\mathrm{dec}}[l]$. Then

$$Z^{\mathrm{dec}} = \bigcup_{j=1}^{[g/2]} Z_j^{\mathrm{dec}}$$

where Z_j^{dec} is the intersection of Z with the image of $\mathcal{A}_j[l] \times \mathcal{A}_{g-j}[l]$ in $\mathcal{A}_g[l]$. Set

$$\overset{\circ}{Z}_j^{\mathrm{dec}} = Z_j^{\mathrm{dec}} - \bigcup_{i \neq j} Z_i^{\mathrm{dec}}.$$

---

[3] This map is generically one-to-one if $g' \neq g''$ and generically two-to-one if $g' = g''$.

Corollary 6.6. *Suppose that $g \geq 4$ and $1 \leq j \leq g/2$ and that Z is a complex submanifold $\overset{\circ}{}$ of $\mathcal{A}_g[l]$ that is contained in $\mathcal{J}_g^*[l]$. If Z is not contained in $\mathcal{J}^{dec}$ and $j \neq 2$, then $\overset{\circ}{Z}_j^{dec}$ is empty or is smooth and has pure codimension 1 in Z. If $\overset{\circ}{Z}_2^{dec}$ is non-empty, it has codimension at most 2 in Z.*

Proof. As in the proof above, it follows from the fact that the image of δ contains no lines in $\mathbb{P}(V' \otimes V'')$ except when $j = 2$. This follows from the fact that canonical curves contain no lines except in genus 2. □

Remark 6.7. One can find further restrictions on the largest strata of Z^{dec}, but so far I have not been able to find a use for them. Partial results in this direction have been obtained in [3] and [4].

7 Locally Symmetric Families of Curves

Definition 7.1. A *locally symmetric family of curves* is a family of stable curves $p : C \to X$ where

(i) X is a locally symmetric variety;

(ii) the Picard group Pic^0 of each curve in the family is an abelian variety (i.e., the dual graph of each fiber is a tree);

(iii) the period map $X \to \mathcal{A}_g$ is a map of locally symmetric varieties (i.e., it is induced by a $\mathbb{Q}$-algebraic group homomorphism).

We lose no generality by assuming that the base X is a simple locally symmetric variety.

By (4.3), every locally symmetric family of curves with all fibers smooth has a ball quotient as base. It becomes more difficult to understand locally symmetric families when the generic fiber is smooth, but some fibers are singular. Let X_j^{dec} be the closure of the locus in X where the fiber has two irreducible components, one of genus j, the other of genus $g - j$. We will assume that $j \leq g/2$. Set

$$\overset{\circ}{X}_j^{dec} = X_j^{dec} - \bigcup_{i \neq j} X_i^{dec}.$$

Proposition 7.2. *Suppose that X is simple and that $C \to X$ is a locally symmetric family of curves of genus g over X whose generic fiber is smooth. If $g \geq 4$, then each X_j^{dec} is a locally symmetric subvariety of X. Moreover, X_j^{dec} is a locally symmetric hypersurface when $j \neq 2$, and $\overset{\circ}{X}_2^{dec}$ has codimension at most 2.*

Proof. Since X is simple, the period map $X \to \mathcal{A}_g$ is constant or finite onto its image. If it is constant, the result is immediate, so we assume that it is finite. Note that X^{dec} is just the preimage of $\mathcal{A}_g^{dec}$, the locus of reducible principally polarized abelian varieties. Since the period map $X \to \mathcal{A}_g$ is a map of locally symmetric varieties, each X_j^{dec} is a locally symmetric subvariety of X. The final statement follows from (6.6). □

Proof of Theorem 1. The theorem is easily seen to be true when $g \leq 2$ as all simple, locally symmetric varieties of complex dimension ≤ 3 and real rank ≤ 2 are ball quotients. Thus we can assume that $g \geq 3$. Fix a level $l \geq 3$ so that $\mathcal{A}_g[l]$ is smooth. By replacing X by a finite covering, we may assume that X is smooth and that the family $C \to X$ is a family of curves with a level l structure, so that we have a period map $X \to \mathcal{A}_g[l]$. The fundamental group Γ of X is a torsion free arithmetic group in a simple $\mathbb{Q}$-group G. Since the period map is not constant, the homomorphism $\Gamma \to Sp_g(\mathbb{Z})$ induced by it has infinite image. (If the image were finite, the family would be isotrivial.)

If all fibers of $p : C \to X$ are smooth, then we have a lift of the homomorphism $\Gamma \to Sp_g(\mathbb{Z})$ to a homomorphism $\Gamma \to \Gamma_g$. If X is good, X^{dec} has codimension ≥ 2, so that the inclusion of $X - X^{\mathrm{dec}}$ into X induces an isomorphism $\pi_1(X - X^{\mathrm{dec}}, *) \cong \Gamma$. Since we have a map $X - X^{\mathrm{dec}} \to \mathcal{M}_g[l]$, we have a homomorphism $\Gamma \to \Gamma_g$ which lifts the homomorphism induced by the period map. Thus in both cases, we have a lift $\Gamma \to \Gamma_g$ of the homomorphism induced by the period map.

Since the image of Γ in $Sp_g(\mathbb{Z})$ is not finite, Theorem 3 implies that the corresponding real group $G(\mathbb{R})$ is isogenous to the product of $SU(n,1)$ with a compact group. It follows that the symmetric space $G(\mathbb{R})/K$ of X is the complex n-ball. $\qquad\square$

8   Locally Symmetric Families of Jacobians

Definition 8.1. A *locally symmetric family of abelian varieties* is a family $B \to X$ of principally polarized abelian varieties where the corresponding map $X \to \mathcal{A}_g$ is a map of locally symmetric varieties. We will call such a family *essential* when the period map is generically finite (and therefore finite) onto its image.

The universal family of abelian varieties over $\mathcal{A}_g[l]$ is essential. If X is simple, then every locally symmetric family of abelian varieties over X is either essential or constant.

Definition 8.2. A *locally symmetric family of jacobians* is a locally symmetric family $B \to X$ of abelian varieties where the image of $B \to \mathcal{A}_g$ lies in the jacobian locus $\mathcal{J}_g$. It is *essential* when its family of jacobians is essential.

The family of jacobians of a locally symmetric family of curves is also a locally symmetric family of jacobians. It is natural to ask whether every locally symmetric family of jacobians can be lifted to a (necessarily locally symmetric) family of curves. The answer is no. The problem is that the period map $\mathcal{M}_g[l] \to \mathcal{A}_g[l]$ is $2:1$ and ramified along the hyperelliptic locus when $g \geq 3$.

Proposition 8.3. *Suppose that $l \geq 3$. There is no family of curves over $\mathcal{A}_3[l]$ whose family of jacobians is the universal family of abelian varieties over $\mathcal{A}_3[l]$.*

Proof. Let $U = \mathcal{A}_3[l] - \mathcal{A}_3^{\mathrm{dec}}[l]$. If the universal family of abelian varieties over $\mathcal{M}_3[l]$ could be lifted to a family of curves, then one would have a section of the period map $\mathcal{M}_3[l] \to U$. This is impossible. One can explain this in two different ways:

(i) When $l \geq 3$, the period map $\mathcal{M}_3[l] \to U$ is 2:1 and ramified along the hyperelliptic locus. So it can have no section.

(ii) If there were such a section, one would have a splitting of the group homomorphism $\Gamma_3[l] \to \pi_1(U)$. But since each component of $\mathcal{A}_3^{\mathrm{dec}}$ has codimension ≥ 2 in $\mathcal{A}_3[l]$, it follows that $\pi_1(U)$ is isomorphic to $\pi_1(\mathcal{A}_3[l])$, which is isomorphic to $Sp_g(\mathbb{Z})[l]$. The homomorphism $\Gamma_3[l] \to \pi_1(U)$ does not split by (3.2).

$\square$

Suppose that $C \to X$ is an essential locally symmetric family of jacobians (or curves). Set $X^* = X \cap \mathcal{J}_g^*[l]$. Let $X_{\mathcal{H}}$ be the locus of points in X where the fiber is hyperelliptic, and let $X_{\mathcal{H}}^* = X^* \cap X_{\mathcal{H}}$.

Proof of Theorem 2. The theorem is easily seen to be true when $g \leq 2$. So we suppose that $g \geq 3$. By passing to a finite index subgroup if necessary, we may suppose that the arithmetic group Γ associated with X is torsion free and that the period map induces a homomorphism $\Gamma \to Sp_g(\mathbb{Z})[l]$ where $l \geq 3$. With these assumptions, we have a period map $X \to \mathcal{A}_g[l]$ and both X and $\mathcal{A}_g[l]$ are smooth. Since X is smooth and good, the inclusion $X^* \hookrightarrow X$ induces an isomorphism on fundamental groups. Since Γ is torsion free, $\pi_1(X, *) \cong \Gamma$.

Now, if X^* is contained in the locus of hyperelliptic jacobians, or disjoint from it, then the period map $X \to \mathcal{A}_g[l]$ has a lift to a map $X \to \mathcal{M}_g[l]$. That is, we can lift the family of jacobians over X^* to a family of curves. This implies that the homomorphism $\Gamma \to Sp_g(\mathbb{Z})$ induced by the period map, lifts to a homomorphism $\Gamma \to \Gamma_g$. Since the period map is not constant, the image of Γ in $Sp_g(\mathbb{Z})$ is not finite (otherwise the family of abelian varieties would be isotrivial). Applying Theorem 3, we see that X is a quotient of the complex n-ball.

The remaining case is where X^* is neither disjoint from the locus of hyperelliptic jacobians nor contained in it. When $g = 3$, the locus of hyperelliptic curves is a divisor in $\mathcal{M}_g[l]$ from which it follows that $X_{\mathcal{H}}^*$ is of pure codimension 1 in X^*. When $g > 3$, it follows from (6.3) that $X_{\mathcal{H}}^*$ is smooth of pure codimension 1 in X^*. Let Y be the fibered product

$$Y = X \times_{\mathcal{A}_g[l]} \mathcal{M}_g[l]$$

The map $Y \to X$ is 2:1 and ramified above the divisor $X_{\mathcal{H}}^*$ in X^*. This implies that $Y \to X^*$ cannot be the restriction of a covering of locally symmetric varieties as such coverings of X are unramified as Γ is torsion free. $\square$

Remark 8.4. To try to eliminate the case where $X_{\mathcal{H}}$ is non-empty, one should study the second fundamental form (with respect to the canonical metric of Siegel space) of the hyperelliptic locus in $\mathcal{A}_g[l]$ and also that of its normal cone. If one is lucky, this will give an upper bound on $\dim X_{\mathcal{H}}$, and therefore of $\dim X$ as well.

9   Appendix: An Example

Suppose that
$$1 \to \mathbb{Z} \to G \to \Gamma \to 1$$
is a non-trivial central extension of a finite index subgroup Γ of $Sp_g(\mathbb{Z})$ by $\mathbb{Z}$. Suppose that the class of this extension is non-trivial in $H^2(\Gamma,\mathbb{Q})$. One can ask whether the natural homomorphism $G \to Sp_g(\mathbb{Z})$ lifts to a homomorphism $G \to \Gamma_g$. We first show that the obstructions of Theorem 3.1 vanish when $g \geq 3$.

Since $g \geq 3$, $H^2(\Gamma,\mathbb{Q}) = \mathbb{Q}$. Since the class of the extension is non-trivial, an easy spectral sequence argument shows that $H^2(G,\mathbb{Q})$ vanishes. Raghunathan's work [21] implies that $H^1(\Gamma,V_3)$ vanishes where V_3 is the third fundamental representation of $Sp_g(\mathbb{R})$. Another easy spectral sequence argument implies that $H^1(G,V_3)$ vanishes. The obstruction to a lift $G \to \Gamma_g$ given in Theorem 3.1 therefore vanishes.

Theorem 9.1. *There is an integer $g_0 \leq 7$ such that if $g \geq g_0$, there is no lift of the natural homomorphism $p : G \to Sp_g(\mathbb{Z})$ to a homomorphism $G \to \Gamma_g$.*

Proof. Suppose that the homomorphism $G \to \Gamma$ lifts to a homomorphism $\phi : G \to \Gamma_g$. The latter homomorphism induces maps on cohomology
$$H^\bullet(\Gamma_g, V) \to H^\bullet(G, V)$$
for all irreducible representations V of $Sp_g(\mathbb{R})$. The cohomology group $H^1(\Gamma_g, V_3)$ is isomorphic to $\mathbb{R}$. For each $Sp_g(\mathbb{R})$ equivariant map $p : V_3^{\otimes 6} \to \mathbb{R}$, there is a natural map
$$p_* : H^1(\Gamma_g, V_3)^{\otimes 6} \to H^6(\Gamma_g, \mathbb{R}).$$
As is traditional, we denote the ith Chern class of the Hodge bundle over $\mathcal{A}_g$ by λ_i. It follows from the work of Kawazumi and Morita [16], that for a suitable choice of p, the image of p_* is spanned by λ_3.

Since the diagram

$$
\begin{array}{ccc}
H^1(\Gamma_g, V_3)^{\otimes 6} & \longrightarrow & H^6(\Gamma_g, \mathbb{R}) \\
\downarrow & & \downarrow{\scriptstyle \phi^\bullet} \\
H^1(G, V_3)^{\otimes 6} & \longrightarrow & H^6(G, \mathbb{R})
\end{array}
$$

commutes, and since $H^1(G, V_3)$ vanishes when $g \geq 3$, it follows that $\phi^* \lambda_3$ is trivial in $H^6(G, \mathbb{R})$. We show that this leads to a contradiction.

It follows from the work of Borel [2] that the ring homomorphism

$$\mathbb{R}[\lambda_1,\lambda_3,\lambda_5,\dots] \to H^\bullet(\Gamma,\mathbb{R})$$

is an isomorphism in degrees $< g$. Since the characteristic class of the extension G of Γ by $\mathbb{Z}$ is a non-zero multiple of λ_1, it follows that the ring homomorphism

$$\mathbb{R}[\lambda_3,\lambda_5,\dots] \to H^\bullet(G,\mathbb{R})$$

is an isomorphism in degrees $< g$. In particular, λ_3 is not zero in $H^6(G,\mathbb{R})$ when $g \geq 7$. It follows that there is no lifting $\phi : G \to \Gamma_g$ when $g \geq 7$. $\qquad\square$

Acknowledgements

This work was supported in part by a grant from the National Science Foundation.

Bibliography

[1] W. Baily, A. Borel: *Compactifications of arithmetic quotients of bounded symmetric domains*, Ann. of Math. (2) 84 (1966), 442–528.

[2] A. Borel: *Stable real cohomology of arithmetic groups*, Ann. Sci. Ecole Norm. Sup. 7 (1974), 235–272.

[3] C. Ciliberto, G. van der Geer: *Subvarieties of the moduli space of curves parameterizing Jacobians with nontrivial endomorphisms*, Amer. J. Math. 114 (1992), 551–570.

[4] C. Ciliberto, G. van der Geer, M. Teixidor i Bigas: *On the number of parameters of curves whose Jacobians possess nontrivial endomorphisms*, J. Algebraic Geom. 1 (1992), 215–229.

[5] R. Coleman: *Torsion points on curves*, in *Galois Representations and Arithmetic Algebraic Geometry*, (Y. Ihara, editor), Advanced Studies in Pure Mathematics 12 (1987), 235–247.

[6] B. Farb, H. Masur: *Superrigidity and mapping class groups*, preprint 1997.

[7] J. Fay: *Theta Functions and Riemann Surfaces*, Lecture Notes in Mathematics no. 352, Springer-Verlag, 1973.

[8] R. Hain: *Completions of mapping class groups and the cycle $C - C^-$*, in *Mapping Class Groups and Moduli Spaces of Riemann Surfaces*, C.-F. Bödigheimer and R. Hain, editors, Contemp. Math. 150 (1993), 75–105.

[9] R. Hain: *Torelli groups and Geometry of Moduli Spaces of Curves*, in *Current Topics in Complex Algebraic Geometry*, C. H. Clemens and J. Kollár, editors, MSRI publications no. 28, Cambridge University Press, 1995.

[10] R. Hain: *Infinitesimal presentations of the Torelli groups*, J. Amer. Math. Soc. 10 (1997), 597–651.

[11] J. Harer: *The second homology group of the mapping class group of an orientable surface*, Invent. Math. 72 (1983), 221–239.

[12] S. Helgason: *Differential Geometry, Lie Groups, and Symmetric Spaces*, Academic Press, 1978.

[13] J. Humphreys: *Introduction to Lie Algebras and Representation Theory*, Graduate Texts in Mathematics 9, Springer-Verlag, 1972.

[14] D. Johnson: *The structure of the Torelli group—III: The abelianization of $\mathcal{I}$*, Topology 24 (1985), 127–144.

[15] J. de Jong, R. Noot: *Jacobians with complex multiplication*, in *Arithmetic algebraic geometry*, G. van der Geer, F. Oort and J. Steenbrink, editors, Progress in Mathematics 89, Birkhäuser, Boston, 1991, 177–192.

[16] N. Kawazumi, S. Morita: *The primary approximation to the cohomology of the moduli space of curves and cocycles for the stable characteristic classes*, Math. Research Letters 3 (1996), 629–641.

[17] G. Margulis: *Discrete subgroups of semisimple Lie groups*, Springer-Verlag, New York, 1991.

[18] S. Morita: *Characteristic classes of surface bundles*, Invent. Math. 90 (1987), 551–577.

[19] D. Mumford: *Towards an enumerative geometry of the moduli space of curves*, in Arithmetic and Geometry, M. Artin and J. Tate, editors (1983), Birkhäuser Verlag (Basel) 271–328.

[20] F. Oort, J. Steenbrink: *The local Torelli problem for algebraic curves*, in *Journées de Géometrie Algébrique d'Angers, Juillet 1979*, edited by A. Beauville, Sijthoff & Noordhoff, Alphen aan den Rijn, 157–204.

[21] M. Raghunathan: *Cohomology of arithmetic subgroups of algebraic groups: I*, Ann. of Math. (2) 86 (1967), 409–424.

[22] I. Satake: *Algebraic Structures of Symmetric Domains*, Mathematics Society of Japan and Princeton University Press, 1980.

[23] J. Tits: *Classification of Algebraic Semisimple Groups*, A. Borel and G.D. Mostow, editors, Proc. Symp. Pure Math. 9 (1966), 33–62.

[24] N. Wallach: Personal communication, fall 1995.

[25] J. Yang: *On the real cohomology of arithmetic groups and the rank conjecture for number fields*, Ann. Scient. Ec. Norm. Sup., 4e série, t. 25, (1992), 287–306.

A Conjectural Description of the Tautological Ring of the Moduli Space of Curves

Carel Faber

Abstract

We formulate a number of conjectures giving a rather complete description of the tautological ring of $\mathcal{M}_g$ and we discuss the evidence for these conjectures.

1 Introduction

Denote by $\mathcal{M}_g$ the moduli space of smooth curves of genus $g \geq 2$ over an algebraically closed field (of arbitrary characteristic). For every $m \geq 3$ that is not divisible by the characteristic, this space is the quotient by a finite group of the nonsingular moduli space of smooth curves with a symplectic level-m structure. Hence the Chow ring $A^*(\mathcal{M}_g)$ can be defined easily—we are working with $\mathbb{Q}$-coefficients throughout this paper. (Cf. [17], Example 8.3.12.)

Some other relevant spaces are the moduli spaces $\mathcal{M}_{g,n}$ of smooth n-pointed curves $(C; x_1, \ldots, x_n)$ of genus g, with $x_i \neq x_j$ for $i \neq j$, defined whenever $2g - 2 + n > 0$, and, denoting $\mathcal{M}_{g,1}$ by $\mathcal{C}_g$, the spaces $\mathcal{C}_g^n$, the n-fold fiber products of $\mathcal{C}_g$ over $\mathcal{M}_g$, parameterizing smooth curves of genus g with n-tuples of not necessarily distinct points. For all these spaces, the Chow ring can be defined as above.

Between these spaces, there are many natural morphisms forgetting one or more points; these will usually be denoted π. The most important of these is $\pi : \mathcal{C}_g \to \mathcal{M}_g$; its relative dualizing sheaf ω_π will often be denoted by ω. This is a $\mathbb{Q}$-line bundle on $\mathcal{C}_g$, cf. [29], p. 299. Writing $K = c_1(\omega) \in A^1(\mathcal{C}_g)$ we define (following Mumford)

$$\kappa_i := \pi_*(K^{i+1}) \in A^i(\mathcal{M}_g)$$

using the ring structure in $A^*(\mathcal{C}_g)$ and the proper push-forward for Chow groups. Note that $\kappa_0 = 2g - 2$ and $\kappa_{-1} = 0$.

Another way to produce natural classes in $A^*(\mathcal{M}_g)$ is to use the Hodge bundle $\mathbb{E} = \pi_*\omega$, a locally free Q-sheaf of rank g on $\mathcal{M}_g$. It is the pull-back of a bundle on $\mathcal{A}_g$, the moduli space of principally polarized abelian varieties of dimension g, via the morphism $t : \mathcal{M}_g \to \mathcal{A}_g$ sending a curve to its Jacobian. We follow Mumford in writing

$$\lambda_i := c_i(\mathbb{E}) \in A^i(\mathcal{M}_g).$$

(Thus $\lambda_0 = 1$ and $\lambda_i = 0$ for $i > g$; the divisor class λ_1 is often denoted λ.)

The κ_i and λ_i are called the *tautological classes*; we define the *tautological ring* of $\mathcal{M}_g$ to be the $\mathbb{Q}$-subalgebra of $A^*(\mathcal{M}_g)$ generated by the tautological classes κ_i and λ_i. We denote this ring by $R^*(\mathcal{M}_g)$.

It can be shown that the classes of many geometrically defined subvarieties of $\mathcal{M}_g$ lie in $R^*(\mathcal{M}_g)$. Examples of subvarieties defined in terms of linear systems for which this happens, will play an important role in this paper. Mumford gives examples (in §7 of [29]) of subvarieties parameterizing curves with special Weierstrass points; these are actually special cases of subvarieties defined in terms of linear systems with certain prescribed types of ramification. Working over $\mathbb{C}$, we have Harer's result [20] that $H^2(\mathcal{M}_g)$ is one-dimensional (for $g \geq 3$); since $H^1(\mathcal{M}_g) = 0$, this implies that

$$A^1(\mathcal{M}_g) = R^1(\mathcal{M}_g) \cong \mathbb{Q}$$

for $g \geq 3$ (note that $\kappa_1 = 12\lambda$, cf. [29] p. 306). (Unfortunately, an algebraic proof of Harer's result is not known.)

For $g \leq 5$, the tautological ring $R^*(\mathcal{M}_g)$ is equal to the Chow ring $A^*(\mathcal{M}_g)$, in characteristic 0. This was shown by Mumford for $g = 2$ (cf. [29] p. 318; it is an immediate consequence of Igusa's description of $\mathcal{M}_2$), by the author for genera 3 and 4 [9, 10] and by Izadi for $g = 5$ [24]. However, it doesn't seem possible that this hold for all g. The idea is that the recent result of Pikaart that the cohomology of $\overline{\mathcal{M}}_g$ is not of Tate type for g large ([30], Cor. 4.7) should imply that the Chow groups of $\overline{\mathcal{M}}_g$ over $\mathbb{C}$ don't map injectively to the cohomology groups; here I am assuming that the result of Jannsen for smooth projective varieties over a universal domain ([25], Thm. 3.6, Rem. 3.11) can be extended to varieties like $\overline{\mathcal{M}}_g$ over $\mathbb{C}$. It would seem then that a similar result would hold for the Chow ring of $\mathcal{M}_g$. The question whether $R^*(\mathcal{M}_g)$ and $A^*(\mathcal{M}_g)$ have the same image in $H^*(\mathcal{M}_g)$ appears to be open.

2   Known Results

We now formulate some known results about the tautological ring of $\mathcal{M}_g$. First of all, Mumford shows ([29], §§5, 6) that the ring $R^*(\mathcal{M}_g)$ is generated by the $g - 2$ classes $\kappa_1, \ldots, \kappa_{g-2}$. The proof has 2 ingredients:

a. Applying the Grothendieck-Riemann-Roch theorem to $\pi : \mathcal{C}_g \to \mathcal{M}_g$ and ω_π gives an expression for the Chern character of the Hodge bundle $\mathbb{E}$ in terms of the κ_i. The resulting expressions for the Chern classes λ_i of $\mathbb{E}$ in the κ_i can be concisely formulated as an identity of formal power series in t:

$$\sum_{i=0}^{\infty} \lambda_i t^i = \exp\left(\sum_{i=1}^{\infty} \frac{B_{2i}\,\kappa_{2i-1}}{2i(2i-1)}\, t^{2i-1}\right).$$

Here B_{2i} are the Bernoulli numbers with signs ($B_2 = 1/6$, $B_4 = -1/30$, $\ldots$). For example

$$\lambda_1 = \frac{1}{12}\kappa_1, \qquad \lambda_2 = \frac{1}{2}\lambda_1^2 = \frac{1}{288}\kappa_1^2, \qquad \lambda_3 = \frac{1}{6}\left(\frac{\kappa_1}{12}\right)^3 - \frac{1}{360}\kappa_3.$$

So all the λ_i can be expressed in the odd κ_i. (Also, the odd κ_i with $i > g$ can be expressed in the lower (odd) kappa's; this will not be used in the proof that $\kappa_1, \ldots, \kappa_{g-2}$ generate, therefore it gives relations between the latter classes in odd degrees greater than g.)

b. On a nonsingular curve, the relative dualizing sheaf is generated by its global sections. This may be formulated universally as the surjectivity of the natural map $\pi^* \mathbb{E} \to \omega$ of locally free sheaves on $\mathcal{C}_g$. The kernel is then locally free of rank $g - 1$ so that its Chern classes vanish in degrees greater than $g - 1$. Hence

$$c_j(\pi^* \mathbb{E} - \omega) = 0 \qquad \forall j \geq g,$$

the difference being taken in the Grothendieck group. Pushing-down to $\mathcal{M}_g$ gives relations between the lambda's and the kappa's in every degree $\geq g - 1$.

To obtain the desired result, one needs to check that in degrees $g - 1$ and g the two relations are independent. For this, Mumford uses an estimate on the size of the Bernoulli numbers; alternatively, one can use (easy) congruence properties of these numbers.

From now on, when talking about relations in the tautological ring, we will mean relations between the kappa's.

Recently, Looijenga proved a strong vanishing result about the tautological ring [27]:

Theorem 1. $R^j(\mathcal{M}_g) = 0$ for all $j > g - 2$ and $R^{g-2}(\mathcal{M}_g)$ is at most one-dimensional, generated by the class of the hyperelliptic locus.

In fact he proved a similar statement for the tautological ring of $\mathcal{C}_g^n$. This is the subring of $A^*(\mathcal{C}_g^n)$ generated by the divisor classes $K_i := pr_i^* K$ and D_{ij} (the class of the diagonal $x_i = x_j$) and the pull-backs from $\mathcal{M}_g$ of the κ_i. The result is that it vanishes in degrees greater than $g - 2 + n$ and that it is at most one-dimensional in degree $g - 2 + n$, generated by the class of the locus

$$\mathcal{H}_g^n = \{(C; x_1, \ldots, x_n) : C \text{ hyperelliptic}; x_1 = \cdots = x_n = x, \text{ a Weierstrass point}\}.$$

Looijenga also gives a description of the degree d part of the tautological ring of $\mathcal{C}_g^n$, for all d and n.

This establishes an important part of one of the conjectures to be discussed in this paper. That conjecture immediately implies Diaz's theorem [5], which gives the upper bound $g - 2$ for the dimension of a complete subvariety of $\mathcal{M}_g$ (in char. 0). Diaz used a flag of subvarieties of $\mathcal{M}_g$ that refines the flag introduced by Arbarello [2]. By using another refinement of Arbarello's flag and by simplifying parts of Diaz's work, Looijenga is able to express every tautological class of degree d as a linear combination of the classes of the irreducible components of a specific geometrically defined locus of n-pointed curves. For $d > g - 2 + n$, the locus is empty; hence the vanishing. In degree $g - 2 + n$, the locus is not irreducible. After having described the irreducible components, Looijenga proves that their classes are proportional to that of $\mathcal{H}_g^n$ by invoking the Fourier transform for abelian varieties (work of Mukai, Beauville and Deninger-Murre). As a corollary of Looijenga's theorem one obtains Diaz's result in arbitrary characteristic.

The question whether the class of the hyperelliptic locus $\mathcal{H}_g$ is actually nonzero was left open; this had been established only for $g = 3$ (due to the existence of complete curves in $\mathcal{M}_3$) and $g = 4$ (by means of a long calculation with 'test surfaces' in $\overline{\mathcal{M}}_4$ (see [10]). Note that the vanishing of $[\mathcal{H}_g]$ would imply an improvement of Diaz's bound; or conversely, the existence of a complete subvariety of dimension $g - 2$ of $\mathcal{M}_g$ would imply the non-vanishing of κ_1^{g-2} (since κ_1 is ample), hence that of $[\mathcal{H}_g]$. However, it is not even known whether $\mathcal{M}_4$ contains a complete surface.

Happily enough, we don't need the existence result for complete subvarieties to settle the non-vanishing in the top degree [12]:

Theorem 2. $\kappa_{g-2} \neq 0$ on $\mathcal{M}_g$. Hence $R^{g-2}(\mathcal{M}_g)$ is one-dimensional.

So the classes $[\mathcal{H}_g]$ and $[\mathcal{H}_g^n]$ are nonzero as well. Before discussing the proof, we remind the reader of the fact that the classes κ_i and λ_i, which so far have been defined only on $\mathcal{M}_g$, can be defined exactly as above on the Deligne-Mumford compactification $\overline{\mathcal{M}}_g$. See [29], §4.

The proof consists of two parts:

a. The class $\lambda_{g-1}\lambda_g$ vanishes on the boundary $\overline{\mathcal{M}}_g - \mathcal{M}_g$ of the moduli space. This is a simple observation. In fact, over Δ_0, the closure of the locus of irreducible singular curves, the class λ_g already vanishes: when pulled back to $\overline{\mathcal{M}}_{g-1,2}$, the Hodge bundle on Δ_0 becomes an extension of a trivial line bundle by the Hodge bundle in genus $g-1$, so its top Chern class vanishes. Over a boundary component Δ_i, with $1 \leq i \leq [g/2]$, the closure of the locus of reducible singular curves consisting of one component of genus i and one of genus $g - i$, the Hodge bundle becomes the direct sum of the Hodge bundles in genera i and $g - i$. Now use the identity $\lambda_h^2 = 0$, valid in arbitrary genus h, to conclude the vanishing of $\lambda_{g-1}\lambda_g$ over Δ_i.

An equivalent formulation is:

a'. The class $ch_{2g-1}(\mathbb{E})$ vanishes on the boundary $\overline{\mathcal{M}}_g - \mathcal{M}_g$ of the moduli space.

One may prove this directly, using on the one hand the additivity of the Chern character in exact sequences and on the other hand the vanishing of all the components of degree $\geq 2h$ of the Chern character of the Hodge bundle in genus h, an easy consequence of the vanishing of the even components proved by Mumford in [29], §5. Or one uses the identity

$$\lambda_{g-1}\lambda_g = (-1)^{g-1}(2g-1)! \cdot ch_{2g-1}(\mathbb{E}),$$

another consequence of Mumford's result.

b. On $\overline{\mathcal{M}}_g$ the following identity holds:

$$\kappa_{g-2}\lambda_{g-1}\lambda_g = \frac{|B_{2g}|(g-1)!}{2^g(2g)!} \quad . \tag{1}$$

This is an identity of intersection numbers (more precisely, the number on the right is the degree of the zero cycle on the left). As B_{2g} doesn't vanish, this proves the theorem.

The first step in the proof of (1) is to use (at last!) the full force of Mumford's result in §5 of [29]: his expression for the Chern character of the Hodge bundle on $\overline{\mathcal{M}}_g$, derived by using the Grothendieck-Riemann-Roch theorem twice. Using this, (1) translates into the following identity of intersection numbers of Witten's tau-classes [32]:

$$\frac{g!}{2^{g-1}(2g)!} = \langle \tau_{g-1}\tau_{2g} \rangle - \langle \tau_{3g-2} \rangle + \frac{1}{2} \sum_{j=0}^{2g-2} (-1)^j \langle \tau_{2g-2-j}\tau_j\tau_{g-1} \rangle$$

$$+ \frac{1}{2} \sum_{h=1}^{g-1} \left((-1)^{g-h} \langle \tau_{3h-g}\tau_{g-1} \rangle \langle \tau_{3(g-h)-2} \rangle + (-1)^h \langle \tau_{3h-2} \rangle \langle \tau_{3(g-h)-g}\tau_{g-1} \rangle \right).$$

The second step is to invoke the Witten conjecture, proven by Kontsevich ([32, 26]). This gives a recipe to compute all intersection numbers of tau-classes: a generating function encoding all these numbers satisfies the Korteweg-de Vries equations. Together with the so-called string and dilaton equations, this determines these numbers recursively.

Proving an explicit identity as the one above, is however not necessarily straightforward from the said recipe. Define the *n-point function* as the following formal power series in n variables $x_1, \ldots, x_n$:

$$\langle \tau(x_1) \cdots \tau(x_n) \rangle = \sum_{a_1,\ldots,a_n \geq 0} \langle \tau_{a_1} \cdots \tau_{a_n} \rangle x_1^{a_1} \cdots x_n^{a_n},$$

which encodes all intersection numbers of n tau-classes. To prove the identity, it suffices to know two special 3-point functions explicitly, to wit $\langle \tau_0 \tau(x) \tau(y) \rangle$ and $\langle \tau(x) \tau(y) \tau(-y) \rangle$. The KdV-equations in the form given by Witten ([32], (2.33)) can be translated into simple differential equations for these functions; an initial condition is provided by the identity

$$\langle \tau_0 \tau_0 \tau(x) \rangle = \exp\left(\frac{x^3}{24} \right) ,$$

which is an immediate consequence of the Witten conjecture. In this way, the two special 3-point functions can be determined easily; for instance,

$$\langle \tau_0 \tau(w) \tau(z) \rangle = \exp\left(\frac{(w^3 + z^3)}{24} \right) \sum_{n \geq 0} \frac{n!}{(2n+1)!} \left(\tfrac{1}{2} wz(w+z) \right)^n ,$$

a formula we learned from Dijkgraaf [6]. This finishes the (sketch of the) proof of the theorem. (Zagier has determined the general 3-point function explicitly.)

3   The Conjectures and the Evidence

We now formulate the first conjecture about the tautological ring $R^*(\mathcal{M}_g)$. We choose to state it first in the form in which it was discussed at several occasions, as early as Spring 1993, in particular, before the two theorems above were proved.

Conjecture 1.   *a. The tautological ring $R^*(\mathcal{M}_g)$ is Gorenstein with socle in degree $g - 2$. I.e., it vanishes in degrees $> g - 2$, is 1-dimensional in degree $g - 2$ and, when an isomorphism $R^{g-2}(\mathcal{M}_g) = \mathbb{Q}$ is fixed, the natural pairing*

$$R^i(\mathcal{M}_g) \times R^{g-2-i}(\mathcal{M}_g) \to R^{g-2}(\mathcal{M}_g) = \mathbb{Q}$$

is perfect.

b. The $[g/3]$ classes $\kappa_1, \ldots, \kappa_{[g/3]}$ generate the ring[1], with no relations in degrees $\leq [g/3]$.[2]

c. There exist explicit formulas for the proportionalities in degree $g - 2$, which may be given as follows. We define expressions $\langle \tau_{d_1+1} \tau_{d_2+1} \cdots \tau_{d_k+1} \rangle$, elements of $R^{g-2}(\mathcal{M}_g)$, in two ways, for every partition of $g - 2$ into positive integers $d_1, d_2, \ldots, d_k$; this allows to express every monomial κ_I of degree $g - 2$ (where I is a multi-index) as a multiple of κ_{g-2}.

---

[1] Morita has recently proved this (explicitly) in cohomology, cf. [28].

[2] After formulating the conjecture, I learned of Harer's improved stability result [21] that essentially implies this.

(1)

$$\langle \tau_{d_1+1}\tau_{d_2+1}\cdots\tau_{d_k+1}\rangle = \frac{(2g-3+k)!(2g-1)!!}{(2g-1)!\prod_{j=1}^{k}(2d_j+1)!!}\,\kappa_{g-2}\,.$$

(2)

$$\langle \tau_{d_1+1}\tau_{d_2+1}\cdots\tau_{d_k+1}\rangle = \sum_{\sigma\in\mathfrak{S}_k}\kappa_\sigma\,,$$

where $\kappa_\sigma = \kappa_{|\alpha_1|}\kappa_{|\alpha_2|}\cdots\kappa_{|\alpha_{\nu(\sigma)}|}$ for a decomposition $\sigma = \alpha_1\alpha_2\cdots\alpha_{\nu(\sigma)}$ of the permutation σ in disjoint cycles, including the 1-cycles; finally, $|\alpha|$ is defined as the sum of the elements in the cycle α, where we think of $\mathfrak{S}_k$ as acting on the k-tuples with entries $d_1, d_2, \ldots, d_k$.

Here $(2a-1)!!$ is shorthand for $(2a)!/(2^a a!)$. A few examples will clarify the recipe given in (c) above:

$$\begin{aligned}
\langle \tau_{g-1}\rangle &= \kappa_{g-2}\,; \\
\langle \tau_i\tau_{g-i}\rangle &= \kappa_{i-1}\kappa_{g-i-1}+\kappa_{g-2} = \frac{(2g-1)!!}{(2i-1)!!(2g-2i-1)!!}\,\kappa_{g-2}\,; \\
\langle \tau_{i+1}\tau_{j+1}\tau_{k+1}\rangle &= \kappa_i\kappa_j\kappa_k+\kappa_{i+j}\kappa_k+\kappa_{i+k}\kappa_j+\kappa_{j+k}\kappa_i+2\kappa_{g-2} \\
&\qquad (i+j+k=g-2).
\end{aligned}$$

Let me point out here that the inspiration to look at the *sums* of 'intersection numbers' (multiples of κ_{g-2}) occurring in (2) above, instead of at the numbers themselves, came entirely from the Witten conjecture ([32], see also [23]), as the notation suggests. A direct link with the actual intersection numbers of Witten's tau-classes on the compactified moduli spaces $\overline{\mathcal{M}}_{g,n}$ was not available at the time, however; it is now, via the class $ch_{2g-1}(\mathbb{E})$ mentioned in the sketch of the proof of Theorem 2. The resulting conjectural identity between the latter numbers is:

$$\frac{(2g-3+k)!}{2^{2g-1}(2g-1)!}\cdot\frac{1}{\prod_{j=1}^{k}(2e_j-1)!!} =$$

$$\langle \tau_{e_1}\cdots\tau_{e_k}\tau_{2g}\rangle - \sum_{j=1}^{k}\langle \tau_{e_1}\cdots\tau_{e_{j-1}}\tau_{e_j+2g-1}\tau_{e_{j+1}}\cdots\tau_{e_k}\rangle$$

$$+\frac{1}{2}\sum_{j=0}^{2g-2}(-1)^j\langle \tau_{2g-2-j}\tau_j\tau_{e_1}\cdots\tau_{e_k}\rangle$$

$$+\frac{1}{2}\sum_{\underline{k}=I\amalg J}\sum_{j=0}^{2g-2}(-1)^j\langle \tau_j\prod_{i\in I}\tau_{e_i}\rangle\langle \tau_{2g-2-j}\prod_{i\in J}\tau_{e_i}\rangle$$

where $\underline{k} = \{1,2,\ldots,k\}$ and $\sum_{j=1}^{k}(e_j-1)=g-2$. (We proved that it is compatible with the string and dilaton equations, so $e_j\geq 2$ may be assumed.)[3]

---

[3] Getzler and Pandharipande [18] have shown that this identity is a consequence of a special case of the Virasoro conjecture of Eguchi, Hori and Xiong [8].

Returning to $\mathcal{M}_g$, we can give the proportionality factor for κ_1^{g-2} explicitly:

$$\kappa_1^{g-2} = \frac{1}{g-1} 2^{2g-5} \big((g-2)!\big)^2 \kappa_{g-2}$$

a formula which we had observed 'experimentally' and which was proven instantaneously by Zagier from (c) above.

Note that parts (a) and (c) of Conjecture 1 implicitly determine the dimension of the $\mathbb{Q}$-vector space $R^i(\mathcal{M}_g)$: it is the rank of the $p(i)$ by $p(g-2-i)$ matrix (with p the partition function) whose entries are the 'intersection numbers' r_{IJ} of monomials κ_I of degree i and κ_J of degree $g-2-i$ given by $\kappa_I \kappa_J = r_{IJ} \kappa_{g-2}$. So the second half of part (b) of the conjecture is the claim that this matrix is of maximal rank whenever $3i \leq g$. Unfortunately, we have not been able to derive an explicit formula for the dimension of $R^i(\mathcal{M}_g)$ in this manner. In joint work with Zagier, we found a relatively simple formula that fits with the data obtained so far from computations; this will be discussed later.

Part (a) of the conjecture may be rephrased to say that $R^*(\mathcal{M}_g)$ has the Poincaré Duality property enjoyed by the ring of algebraic cohomology classes (with $\mathbb{Q}$-coefficients) of a nonsingular projective variety of dimension $g-2$. In light of this, Thaddeus asked the question whether $R^*(\mathcal{M}_g)$ also satisfies the other properties such a ring is known (resp. conjectured) to have in char. 0 (resp. in char. $p > 0$). (Cf. Grothendieck's paper [19] for a discussion of these.) After having examined the available evidence, we feel confident enough to extend Conjecture 1:

Conjecture 1(bis). *In addition to the properties mentioned in Conjecture 1, $R^*(\mathcal{M}_g)$ 'behaves like' the algebraic cohomology ring of a nonsingular projective variety of dimension $g-2$; i.e., it satisfies the Hard Lefschetz and Hodge Positivity properties with respect to the class κ_1.*

We don't have a particular candidate for such a projective variety. Diaz's upper bound allows for the existence of such a variety lying inside $\mathcal{M}_g$, although presumably it will have at least quotient singularities in that case. For a brief discussion of the relation between the conjectured form of the tautological ring and the occurrence of complete subvarieties inside moduli space, see the concluding remarks.

So as not to lose the interest of the skeptical reader, we state the following result.

Theorem 3. *Conjectures 1 and 1(bis) are true for all $g \leq 15$.*

In order to explain how we could settle these conjectures for the values of g mentioned, we introduce certain sheaves on the spaces $\mathcal{C}_g^d$, the d-fold fiber products of the universal curve $\mathcal{C}_g$ over $\mathcal{M}_g$.

Consider the projection $\pi = \pi_{\{1,\ldots,d\}} : \mathcal{C}_g^{d+1} \to \mathcal{C}_g^d$ that forgets the $(d+1)$-st point. Denote by Δ_{d+1} the sum of the d divisors $D_{1,d+1}, \ldots, D_{d,d+1}$ as well as (by abuse of notation) its class:

$$\Delta_{d+1} = D_{1,d+1} + \cdots + D_{d,d+1}.$$

Further, denote by ω_i the (Q-)line bundle on C_g^n obtained by pulling back ω on C_g along the projection onto the i-th factor and denote its class in the codimension-1 Chow group by K_i.

We define a coherent sheaf $\mathbb{F}_d$ on C_g^d by the formula

$$\mathbb{F}_d = \pi_*(\mathcal{O}_{\Delta_{d+1}} \otimes \omega_{d+1}).$$

The sheaf $\mathbb{F}_d$ is locally free of rank d; its fiber at a point $(C; x_1, \ldots, x_d) = (C; D)$ is the vector space

$$H^0(C, K/K(-D)).$$

We think of $\mathbb{F}_d$ as a universal d-pointed jet bundle. It is invariant for the action of $\mathfrak{S}_d$. Its total Chern class can be expressed in terms of the tautological divisor classes K_i and D_{ij} on C_g^d:

$$\begin{aligned}
c(\mathbb{F}_d) &= (1 + K_1)(1 + K_2 - \Delta_2)(1 + K_3 - \Delta_3) \cdots (1 + K_d - \Delta_d) \\
&= (1 + K_1)(1 + K_2 - D_{12})(1 + K_3 - D_{13} - D_{23}) \cdots (1 + K_d - D_{1d} \cdots - D_{d-1,d}).
\end{aligned}$$

This can be proved for instance using the Grothendieck-Riemann-Roch theorem and a natural filtration on the sheaves $\mathbb{F}_n$.

The natural evaluation map of locally free sheaves on C_g^d:

$$\varphi_d : \mathbb{E} \to \mathbb{F}_d$$

between the pulled-back Hodge bundle of rank g and the bundle $\mathbb{F}_d$ of rank d will be our main tool in constructing relations between tautological classes. Fiberwise the kernel over $(C; D)$ is the vector space $H^0(C, K(-D))$, whose dimension may vary with D.

Observe that the locus $\{\mathrm{rk}\,\varphi_d \leq d - r\}$ parameterizes the pairs $(C; D)$ for which $\dim H^0(C, K(-D)) \geq g - d + r$, equivalently, $\dim H^0(C, D) \geq r + 1$, in other words, for which the complete linear system $|D|$ has dimension at least r. So the image of this locus in $\mathcal{M}_g$ parameterizes the curves possessing a g_d^r. The expected codimension of the locus $\{\mathrm{rk}\,\varphi_d \leq d - r\}$ in C_g^d is $r(g - d + r)$; the expected fiber dimension of the map to $\mathcal{M}_g$ is r; so the expected codimension in $\mathcal{M}_g$ of the locus of curves possessing a g_d^r is

$$\rho = r(g - d + r) - d + r = (r + 1)(g - d + r) - g,$$

the Brill-Noether number.

Porteous's formula (cf. [1, 17]) computes the class of the degeneracy locus $\{\mathrm{rk}\,\varphi_d \leq d - r\}$ if it is either empty or has the expected codimension. The formula is:

$$\mathrm{class}\,(\{\mathrm{rk}\,\varphi_d \leq d - r\}) = \Delta_{r, g-d+r}(c(\mathbb{F}_d - \mathbb{E})).$$

Here the difference is taken in the Grothendieck group; $c(\mathbb{F}_d - \mathbb{E})$ is the formal power series in t obtained as the quotient $c(\mathbb{F}_d)/c(\mathbb{E})$ of total Chern classes, this time written as polynomials in t. Finally,

$$
\Delta_{p,q}\left(\sum_{i=0}^{\infty} c_i t^i\right) =
\begin{vmatrix}
c_p & c_{p+1} & \cdots & c_{p+q-1} \\
c_{p-1} & c_p & \cdots & c_{p+q-2} \\
\vdots & \vdots & \ddots & \vdots \\
c_{p-q+1} & c_{p-q+2} & \cdots & c_p
\end{vmatrix}.
$$

As a very important example of the above, consider the curves C of genus g with a g_{2g-1}^g. Divisors of degree -1 on a curve are not effective; dually this says that no curve has a g_{2g-1}^g. Hence

$$\{\mathrm{rk}\ \varphi_{2g-1} \leq g - 1\}$$

is a fancy way to denote the empty set. Porteous's formula applies and we find:

Proposition 3.1.

$$c_g(\mathbb{F}_{2g-1} - \mathbb{E}) = 0.$$

As the Chern classes of both $\mathbb{E}$ and $\mathbb{F}_{2g-1}$ are expressed in terms of the tautological classes, this gives a relation between tautological classes on C_g^{2g-1}. In principle, pushing-down this relation all the way to $\mathcal{M}_g$ will give a relation between the tautological classes κ_i and λ_i, hence between the κ_i themselves. (One has to note that a monomial in the classes K_i and D_{ij} pushes down to a monomial in the κ_i; the formulas governing this will be discussed shortly.)

Unfortunately, for trivial reasons the obtained relation is identically zero: the relation of the proposition lives in codimension g on C_g^{2g-1}, hence ends up in negative codimension once pushed down to $\mathcal{M}_g$. (This might be one of the reasons why this relation apparently was not considered before.)

Fortunately however, "once zero, always zero": multiplying the relation with an arbitrary class gives another relation; in particular, multiplying it with a monomial in tautological classes gives another relation between tautological classes; after pushing-down to $\mathcal{M}_g$ we obtain relations between the κ_i which do not obviously vanish.

Before discussing these relations, we state a variant of the proposition above:

Proposition 3.2.

For all $d \geq 2g - 1$, for all $j \geq d - g + 1$, $c_j(\mathbb{F}_d - \mathbb{E}) = 0.$

This is because the locus $\{\mathrm{rk}\ \varphi_d \leq g - 1\}$ is empty; so φ_d is an injective map of vector bundles, whence the cokernel is locally free of rank $d - g$, whence the result. (To ease the exposition, we will only use $d = 2g - 1$ in the sequel.)

The available evidence suggests that the relations between the κ_i obtained from the relations just stated by multiplying with a monomial in the K_i and D_{ij} and pushing-down to $\mathcal{M}_g$ are very non-trivial indeed. In fact, calculations we have done show that for $g \leq 15$ these relations generate the entire ideal of relations in the tautological ring. I.e., dividing out the polynomial ring $\mathbb{Q}[\kappa_1, \dots, \kappa_{g-2}]$ by the ideal of relations so obtained gives a quotient ring that surjects onto the tautological ring; the quotient ring is Gorenstein with socle in degree $g-2$; because $R^{g-2}(\mathcal{M}_g)$ is nonzero by Theorem 2, the surjection is in fact an isomorphism. In this way one proves Theorem 3. Below we discuss the calculations in some detail. Because we see no reason whatsoever why the result of the calculations would be different for higher genera, we put forward the following conjecture:

Conjecture 2. *In the polynomial ring* $\mathbb{Q}[\kappa_1, \dots, \kappa_{g-2}]$, *let* I_g *be the ideal generated by the relations of the form*

$$\pi_*\big(M \cdot c_j(\mathbb{F}_{2g-1} - \mathbb{E})\big),$$

with $j \geq g$ *and* M *a monomial in the* K_i *and* D_{ij} *and* $\pi : \mathcal{C}_g^{2g-1} \to \mathcal{M}_g$ *the forgetful map. Then the quotient ring* $\mathbb{Q}[\kappa_1, \dots, \kappa_{g-2}]/I_g$ *is Gorenstein with socle in degree* $g - 2$; *hence it is isomorphic to the tautological ring* $R^*(\mathcal{M}_g)$.

The implied isomorphism follows from Theorem 2. As mentioned, Conjecture 2 is proved for all $g \leq 15$. (As it turns out, monomials in the D_{ij} suffice.)

4   Calculations

We now discuss the 'mechanics' of the calculation: how a relation of the form stated in Conjecture 2 actually produces a relation between the κ_i on $\mathcal{M}_g$. Observe that the expression

$$M \cdot c_j(\mathbb{F}_{2g-1} - \mathbb{E})$$

may be expanded into a polynomial in the classes K_i, D_{ij} and λ_i. The latter classes are pull-backs from $\mathcal{M}_g$; we will essentially always suppress this in the notation[4]. Also, although strictly speaking we don't need it at this point, we point out that it is easy to invert $c(\mathbb{E})$ (see [29], §5):

$$c(\mathbb{E})^{-1} = c(\mathbb{E}^\vee) = 1 - \lambda_1 + \lambda_2 - \lambda_3 + \cdots + (-1)^g \lambda_g.$$

Hence it suffices to explain how to compute π_* of a monomial in the classes just mentioned. The projection formula tells us that it suffices to do this for monomials in the classes K_i and D_{ij}. Now the map π is the composition of morphisms π_d from $\mathcal{C}_g^d$ to $\mathcal{C}_g^{d-1}$ forgetting the d-th point:

---

[4] Similarly for the kappa's in what follows.

$$\pi = \pi_1 \circ \pi_2 \circ \cdots \circ \pi_{2g-1} \,.$$

The formulas for computing $\pi_{d,*}$ of a monomial were already stated in [22], top of p. 55:

Formularium.

a. *Every monomial in the classes K_i $(1 \le i \le d)$ and D_{ij} $(1 \le i < j \le d)$ on C_g^d can be rewritten as a monomial M pulled back from C_g^{d-1} times either a single diagonal D_{id} or a power K_d^k of K_d by a repeated application of the following substitution rules:*

$$\begin{cases} D_{id} D_{jd} \to D_{ij} D_{id} & (i < j < d); \\ D_{id}^2 \to -K_i D_{id} & (i < d); \\ K_d D_{id} \to K_i D_{id} & (i < d). \end{cases}$$

b. *For M a monomial pulled back from C_g^{d-1} :*

$$\begin{cases} \pi_{d,*}(M \cdot D_{id}) = M; \\ \pi_{d,*}(M \cdot K_d^k) = M \cdot \pi^*(\kappa_{k-1}). \end{cases}$$

Here $\pi : C_g^{d-1} \to \mathcal{M}_g$ is the forgetful map.

Note that $\pi_{d,*}(M) = 0$ as it should be, since $\kappa_{-1} = 0$.

Let us discuss some examples, starting in genus 2. On $\mathcal{C}_2^3$ we have:

$$\begin{aligned} 0 &= c_2(\mathbb{F}_3 - \mathbb{E}) \\ &= c_2\big((1 + K_1)(1 + K_2 - D_{12})(1 + K_3 - D_{13} - D_{23})(1 - \lambda_1 + \lambda_2)\big) \\ &= (K_1 K_2 + K_1 K_3 + K_2 K_3 - K_1 D_{12} - K_1 D_{13} - K_1 D_{23} \\ &\quad - K_2 D_{13} - K_2 D_{23} - K_3 D_{12} + D_{12} D_{13} + D_{12} D_{23}) \\ &\quad - \lambda_1(K_1 + K_2 + K_3 - D_{12} - D_{13} - D_{23}) + \lambda_2 \,. \end{aligned}$$

Upon intersecting this with D_{13} and applying the substitution rules, this becomes:

$$\begin{aligned} 0 &= D_{13} \cdot c_2(\mathbb{F}_3 - \mathbb{E}) \\ &= (2K_1^2 D_{13} + 3K_1 K_2 D_{13} - 6K_1 D_{12} D_{13}) \\ &\quad - \lambda_1(3K_1 D_{13} + K_2 D_{13} - 2D_{12} D_{13}) + \lambda_2 D_{13} \,. \end{aligned}$$

Pushing this down gives:

$$\begin{aligned} 0 &= \pi_*\big(D_{13} \cdot c_2(\mathbb{F}_3 - \mathbb{E})\big) \\ &= (\pi_1 \circ \pi_2)_*\big(2K_1^2 + 3K_1 K_2 - 6K_1 D_{12} - \lambda_1(3K_1 + K_2 - 2D_{12}) + \lambda_2\big) \\ &= \pi_{1,*}(3K_1 \kappa_0 - 6K_1 - \lambda_1 \kappa_0 + 2\lambda_1) \\ &= 3\kappa_0^2 - 6\kappa_0 = 0 \,. \end{aligned}$$

This is no surprise. Instead, pushing-down after intersecting with $D_{12}D_{13}$ gives:

$$
\begin{aligned}
0 &= \pi_*\big(D_{12}D_{13} \cdot c_2(\mathbb{F}_3 - \mathbb{E})\big) \\
&= (\pi_1 \circ \pi_2)_*(11K_1^2 D_{12} - 6\lambda_1 K_1 D_{12} + \lambda_2 D_{12}) \\
&= \pi_{1,*}(11K_1^2 - 6\lambda_1 K_1 + \lambda_2) \\
&= 11\kappa_1 - 6\kappa_0\lambda_1 = 11\kappa_1 - 12\lambda_1 = 10\kappa_1 \,,
\end{aligned}
$$

which implies $\kappa_1 = 0$. Although we knew this already, it nevertheless shows that non-trivial relations can be obtained in this manner. In fact, using $\kappa_1 = 0$, hence $\lambda_1 = 0 = \lambda_2$, we obtain also the relations:

$$
\left\{
\begin{aligned}
K_1^2 &= 0; \\
K_1 K_2 &= 2K_1 D_{12}; \\
0 &= K_1 D_{12} + K_1 D_{13} + K_2 D_{23} - K_1 D_{23} - K_2 D_{13} - K_3 D_{12} + 2D_{12}D_{13}.
\end{aligned}
\right.
$$

(The symmetry of the situation was used to obtain the latter relation.) As one can check easily, it follows that $R^*(\mathcal{C}_2^n)$ is Gorenstein (with socle in degree n) for $n \leq 3$.[5]

Next we look at genus 3. We know that $\kappa_1 \neq 0$, so there are no relations in $R^1(\mathcal{M}_3)$. The vanishing in codimension 2 resulting from Looijenga's theorem can be made explicit by means of the relations introduced above. First, upon intersecting $c_3(\mathbb{F}_5 - \mathbb{E})$ with the product $D_{12}D_{13}D_{14}D_{15}$ of diagonals and pushing this down to $\mathcal{C}_3$ we find:

$$
\begin{aligned}
0 &= c_3\big((1 + K)(1 + 2K)\cdots(1 + 5K) - \mathbb{E}\big) \\
&= 225K^3 - 85K^2\lambda_1 + 15K\lambda_2 - \lambda_3 \,,
\end{aligned}
$$

which after pushing-down to $\mathcal{M}_3$ gives the relation

$$
225\kappa_2 - \frac{55}{8}\kappa_1^2 = 0.
$$

Second, intersecting $c_4(\mathbb{F}_5 - \mathbb{E})$ with the product $D_{12}D_{13}D_{45}$ and pushing this down to $\mathcal{M}_3$ one finds (after a calculation) the relation

$$
\frac{87}{4}\kappa_1^2 - 162\kappa_2 = 0.
$$

Hence we find $\kappa_1^2 = 0 = \kappa_2$, so that $R^*(\mathcal{M}_3) = \mathbb{Q}[\kappa_1]/(\kappa_1^2)$, as was already known to Mumford (cf. [29], p. 309).

---

[5] A recent result of Pandharipande and the author [15] implies that this is true for all n.

In genus 4, the interesting codimension is 2: we know that $R^2(\mathcal{M}_4)$ is 1-dimensional; the question is what the precise relation between κ_1^2 and κ_2 is. It was determined in [10]; let us rederive it here. Intersecting $c_4(\mathbb{F}_7 - \mathbb{E})$ with the product $D_{12}D_{13}D_{14}D_{15}D_{67}$ and pushing this down to $\mathcal{M}_4$ we find eventually the relation

$$150\kappa_1^2 - 1600\kappa_2 = 0.$$

Similarly, via $D_{12}D_{13}D_{14}D_{56}D_{57} \cdot c_4(\mathbb{F}_7 - \mathbb{E})$ respectively $D_{12}D_{34}D_{56}D_{57} \cdot c_5(\mathbb{F}_7 - \mathbb{E})$ we find the relations

$$360\kappa_1^2 - 3840\kappa_2 = 0 \quad \text{and} \quad -180\kappa_1^2 + 1920\kappa_2 = 0.$$

Lengthy calculations are required to obtain these relations, especially in the last case. It is then reassuring to find that all three are equivalent to

$$\kappa_1^2 = \frac{32}{3}\kappa_2,$$

as obtained in [10] and in accordance with the predicted relation between κ_1^{g-2} and κ_{g-2} in genus g mentioned before.

This may be a good moment to point out that the relations described in Conjecture 2, geometrically transparent as they are, appear to be rather complicated from a combinatorial point of view. To illustrate this, we observe that $c_g(\mathbb{F}_{2g-1} - \mathbb{E})$ is a polynomial of degree g in roughly $2g^2$ variables. If expanding it completely before applying the substitution and push-down rules of the Formularium were the only option, we would be stuck in genus 5 or 6 already. As we will explain later, there are various ways to overcome this difficulty. For the moment we continue our description of the tautological rings in low genus. All results were obtained by explicitly calculating the relations described in Conjecture 2. We wrote several Maple[6] procedures for the occasion.

Genus 5 gives the first example of a relation not resulting from Looijenga's theorem. It is the relation

$$\kappa_1^2 = \frac{72}{5}\kappa_2,$$

establishing that $R^2(\mathcal{M}_5)$ is 1-dimensional, thus providing the first evidence beyond Theorems 1 and 2 for the conjectured Gorenstein property of $R^*(\mathcal{M}_g)$. We also find the predicted relation $\kappa_1^3 = 288\kappa_3$. Let us also give the resulting relations between kappa's and lambda's:

$$\begin{cases} \kappa_1 = 12\lambda_1 \\ \kappa_2 = 10\lambda_1^2 = 20\lambda_2 \\ \kappa_3 = 6\lambda_1^3 = 40\lambda_3 \end{cases}.$$

Combining these results with the computation of $\kappa_{g-2}\lambda_{g-1}\lambda_g$, we computed the tautological class of the locus of Jacobians of curves of genus 5 in the moduli space $\mathcal{A}_5$ of principally polarized abelian 5-folds[7]. The method is the same as in [11], §5; see also the concluding remarks. The result is:

---

[6] Maple© is a trademark of the University of Waterloo and Waterloo Maple Software.
[7] In [13] we show how to compute this for arbitrary g.

$$[\mathcal{J}_5]_Q = \frac{1}{2}[\mathcal{J}_5] = 36\lambda_1^3 - 48\lambda_3 + X$$

in the rational cohomology of a toroidal compactification $\widetilde{\mathcal{A}}_5$, where X satisfies $\lambda_1^{12}X = \lambda_1^9\lambda_3 X = 0$.

Note that the relations above, together with Theorem 2, determine the Chow ring of $\mathcal{M}_5$, since Izadi proved that the Chow ring equals the tautological ring in genus 5 [24].

We remark that for $g \leq 5$

$$R^*(\mathcal{M}_g) = \mathbb{Q}[\kappa_1]/(\kappa_1^{g-1})$$

(and the tautological ring equals the Chow ring). Such a simple description is not available in genus 6 and higher: κ_1^2 and κ_2 are independent, as follows from Edidin's result [7].

We find in genus 6:

$$R^*(\mathcal{M}_6) = \mathbb{Q}[\kappa_1,\kappa_2]/(127\kappa_1^3 - 2304\,\kappa_1\kappa_2, 113\,\kappa_1^4 - 36864\,\kappa_2^2).$$

Hence this is still a complete intersection ring. We also have the relations

$$\kappa_3 = \frac{5}{2304}\,\kappa_1^3 \quad \text{and} \quad \kappa_4 = \frac{5}{73728}\,\kappa_1^4.$$

In genera 7 and 8, the tautological rings are also complete intersection rings, but apparently this is not the case for any genus greater than or equal to 9. As a final example, we give here the tautological ring in genus 9, because it is the first one which is not a complete intersection, and also to give the reader an idea of how complicated these rings become very quickly. The ring $R^*(\mathcal{M}_9)$ is the quotient of $\mathbb{Q}[\kappa_1,\kappa_2,\kappa_3]$ by the ideal generated by

$$\begin{cases} 5195\,\kappa_1^4 + 3644694\,\kappa_1\kappa_3 + 749412\,\kappa_2^2 - 265788\,\kappa_1^2\kappa_2 \\ 33859814400\,\kappa_2\kappa_3 - 95311440\,\kappa_1^3\kappa_2 + 2288539\,\kappa_1^5 \\ 19151377\,\kappa_1^5 + 16929907200\,\kappa_1\kappa_2^2 - 1142345520\,\kappa_1^3\kappa_2 \\ 1422489600\,\kappa_3^2 - 983\,\kappa_1^6 \\ 1185408000\,\kappa_2^3 - 47543\,\kappa_1^6 \\ 42019\,\kappa_1^6 - 1234800\,\kappa_1^4\kappa_2 \end{cases}$$

(the last generator is actually superfluous). We also have the relations

$$\begin{cases} 1399562496\,\kappa_4 = 2453760\,\kappa_2^2 - 65425\,\kappa_1^4 + 2470320\,\kappa_1^2\kappa_2 \\ 7223427072\,\kappa_5 = 307440\,\kappa_1^3\kappa_2 - 8729\,\kappa_1^5 \\ 309657600\,\kappa_6 = \kappa_1^6 \\ 26011238400\,\kappa_7 = \kappa_1^7 \end{cases}$$

So the dimensions of the vector spaces $R^i(\mathcal{M}_9)$ are 1,1,2,3,3,2,1,1 respectively. By straightforward calculations one verifies, first, that the pairings $R^i(\mathcal{M}_9) \times R^{7-i}(\mathcal{M}_9) \to R^7(\mathcal{M}_9) = \mathbb{Q}$ are perfect, i.e., $R^*(\mathcal{M}_9)$ is Gorenstein, second, that $R^*(\mathcal{M}_9)$ satisfies the Hard Lefschetz and Hodge Positivity properties, third, that the proportionalities in degree $g - 2 = 7$ are as conjectured. All the relations above are in the ideal I_9 mentioned in Conjecture 2. This completes the proof of Conjectures 1 and 2 in the case $g = 9$ at hand.

To prove Conjectures 1 and 2 for other values of g, one proceeds entirely analogously. I would like to point out that settling these two conjectures can be viewed as a combinatorial problem. For a given value of g, the verification of the conjectures amounts to a finite calculation; in principle it can be done on a computer. In practice, we carried this out for $g \leq 15$. As remarked already, this requires an implementation in which the expressions $c_g(\mathbb{F}_{2g-1} - \mathbb{E})$ are *not* expanded completely before the rules of the formularium are applied to them. First, note that every diagonal D_{ij} in M can be used to reduce the number of points by 1, by applying the substitution and push-down rules involving D_{ij} directly to $\mathbb{F}_d$. One finds d-pointed jet bundles on $\mathcal{C}_g^k$ (with $k < d$), where the d points appear with multiplicities at the general point of $\mathcal{C}_g^k$. E.g., a $D_{d-1,d}$ occurring in M has the effect of replacing $(1 + K_d - \Delta_d)$ in $c(\mathbb{F}_d)$ by $(1 + 2K_{d-1} - \Delta_{d-1})$; this gives the total Chern class of the d-pointed jet bundle on $\mathcal{C}_g^{d-1}$ corresponding to d-tuples that contain the $(d-1)$-st point with multiplicity 2. For another example, see the calculation done for $c_3(\mathbb{F}_5 - \mathbb{E})$ in genus 3 above.

Next, the trivial identity

$$c_k(\mathbb{F}_d) = c_k(\mathbb{F}_{d-1}) + (K_d - \Delta_d)\, c_{k-1}(\mathbb{F}_{d-1})$$

can be used to expand the total Chern classes of d-pointed jet bundles (possibly with multiplicities) *step by step*. This makes the computations somewhat more manageable. We remark that $\pi_*\big(M \cdot c_j(\mathbb{F}_{2g-1} - \mathbb{E})\big)$ is trivially 0 if not all indices from $\{1, \ldots, 2g-1\}$ are 'covered' by the monomial M.

Moreover, we derived a formula expressing $\pi_*\big(\Pi_\alpha \cdot c_k(\mathbb{F}_d)\big)$ directly as a polynomial in the kappa's, where $\alpha = (\alpha_1, \ldots, \alpha_a)$ is a partition of d and Π_α is a product of $d - a$ diagonals corresponding to α:

$$\Pi_\alpha = D_{1,2} \cdots D_{1,\alpha_1} D_{\alpha_1+1,\alpha_1+2} \cdots D_{\alpha_1+1,\alpha_1+\alpha_2} \cdots D_{d-\alpha_a+1,d-\alpha_a+2} \cdots D_{d-\alpha_a+1,d}.$$

In joint work with Zagier, this formula was rewritten in a form involving formal power series. This form was then used to prove the following statements about the relations in the ideal I_g introduced in Conjecture 2:

a. I_g contains no relations in codimensions $\leq g/3$.

b. For g of the form $3k - 1$ with k an integer, there is a unique relation in codimension k in I_g.

c. Write $g = 3k - 1 - \ell$, with k and ℓ positive integers. There exists an upper bound for the number of relations in I_g in codimension k, which only depends on ℓ.

The first statement is consistent with the second half of part (b) of Conjecture 1 (but doesn't prove it). As remarked, the improved stability result of Harer [21] essentially implies that there are no relations between the kappa's in codimensions $\leq g/3$.

The unique relation in codimension k and genus $3k - 1$ is given as follows. Define rational numbers a_i for $i \geq 1$ via the identity

$$\exp\left(-\sum_{i=1}^{\infty} a_i t^i\right) = \sum_{n=0}^{\infty} \frac{(6n)!}{(2n)!(3n)!} t^n$$

of formal power series. Then the said relation is the coefficient of t^k in

$$\exp\left(\sum_{i=1}^{\infty} a_i \kappa_i t^i\right).$$

As to the third statement, the available computational evidence suggests that the *actual* number of relations in $R^*(\mathcal{M}_g)$ in codimension k and genus $g = 3k - 1 - \ell$ depends only on ℓ, whenever $2k \leq g-2$ (i.e., $k \geq \ell+3$). Assuming this, and denoting this number by $a(\ell)$, we know that $a(\ell) = 1,1,2,3$ for $\ell = 0,1,2,3$ respectively; further computations assuming Conjecture 1 give the following results:

ℓ	0	1	2	3	4	5	6	7	8	9
$a(\ell)$	1	1	2	3	5	6	10	13	18	24

Zagier and I have a favourite guess as to what the function a might be (but there are many functions with 10 prescribed values).

Detailed results concerning the formula expressing $\pi_*\left(\Pi_\alpha \cdot c_k(\mathbb{F}_d)\right)$ as a polynomial in the kappa's as well as the work with Zagier will appear in [16].

5 Other Relations

So far, we have only considered relations resulting from the triviality that divisors of negative degree on a curve are not effective (or rather from the dual statement). Naturally, many more relations can be produced with the same method. Start with a triple (g,d,r) such that the locus $\{\mathrm{rk}\ \varphi_d \leq d - r\}$ has the expected codimension, so that its class can be computed using Porteous's formula. If the fibers of the map to $\mathcal{M}_g$ have the expected dimension r, then cutting the locus with r sufficiently general divisors will give a locus that maps finitely onto the locus in $\mathcal{M}_g$ of curves possessing a g_d^r. Often, divisors can be chosen whose classes lie in the tautological ring of C_g^d, and often there are quite a few possible choices for such divisors. Every such choice leads to a formula for the class of the locus in $\mathcal{M}_g$ of curves possessing a g_d^r, with a certain multiplicity; if the multiplicity can be computed—this is often the case—we obtain the actual class, as an element of the tautological ring. (These observations are certainly not new; see for instance the papers [4, 22, 29, 31].)

Equating the various formulas for this class leads then to relations in the tautological ring. However, if one is after such relations (we are), it is a lot easier to cut with *fewer* than r divisors whose classes lie in the tautological ring; pushing-down the resulting class to $\mathcal{M}_g$ gives a polynomial in the kappa's which is 0 in the ring, since the fibers of the map to $\mathcal{M}_g$ are positive-dimensional. In this way we obtain relations without having to compute multiplicities.

Examples:

a. 1-dimensional linear systems. The loci involved have the expected dimensions. The locus of divisors moving in a g_d^1 has dimension $2g - 4 + 2d$. Cutting its class with K_1 (i.e., requiring that the first point be in a fixed canonical divisor) gives a class that pushes down to the class of curves with a g_d^1, with multiplicity $(d-1)!(2g-2)$. Cutting it with D_{12} instead, we find that class with multiplicity $(d-2)!(2g+2d-2)$. The two formulas are worked out in [31], §5, for $d = 2$ resp. 3. In [10] we proved that the resulting relation expresses κ_{g-2} (resp. κ_{g-4}) in lower kappa's, for $g \geq 4$ (resp. for $g \geq 7$). The other option is to push-down directly. This produces a non-trivial relation in codimension $g - 2d + 1$. It appears that together these relations are sufficient to prove the first half of part (b) of Conjecture 1, but I have not been able to carry out the calculation.

b. Plane quintics. They form a 12-dimensional subvariety of $\mathcal{M}_6$. Inside C_6^5 the divisors moving in a g_5^2 form a 14-dimensional locus. Note that it has two components: one lies over the plane quintics; the other lies over the hyperelliptic locus, with 3-dimensional fibers (the g_5^2 equals the g_4^2 plus an arbitrary base point). Cutting the class with two divisors and pushing-down, we find the class of the plane quintics in $\mathcal{M}_6$, with a certain multiplicity. There are 5 ways of choosing two tautological divisors: $D_{12}D_{34}$, $D_{12}D_{13}$, K_1D_{23}, K_1D_{12} and K_1K_2. As one checks easily, these give the class of the plane quintics with multiplicities 240, 90, 360, 60 and 600 respectively. (E.g., K_1D_{23} puts the first point in a fixed canonical divisor; the second and third point coincide, so this is a point of tangency on a line through the first point; this fixes the fourth and fifth point, but not their order, for a total of $10 \cdot 18 \cdot 2 = 360$ possibilities.) The 5 formulas for the class give 4 relations between κ_1^3, $\kappa_1\kappa_2$ and κ_3. It turns out these relations have rank 2 (the actual rank, as we know now). (When I did these calculations originally (Summer 1991), the result strongly suggested to me that $R^3(\mathcal{M}_6)$ should be 1-dimensional; this was instrumental in formulating the first version of Conjecture 1 later that year.) The class of the plane quintics turns out to be $\frac{35}{3072}\kappa_1^3$.

c. Hyperelliptic curves. If Porteous's formula can be applied to compute the class of divisors moving in a g_d^r, it can also be applied to compute the class of the divisors moving in the dual system (a $g_{2g-2-d}^{g-d+r-1}$). Let us apply this to hyperelliptic curves. The class of divisors moving in a g_{2g-4}^{g-2} is

$$\Delta_{g-2,2}\big(c(\mathbb{F}_{2g-4} - \mathbb{E})\big) = (c_{g-2}^2 - c_{g-1}c_{g-3})(\mathbb{F}_{2g-4} - \mathbb{E}).$$

This has $(g-2)$-dimensional fibers over the hyperelliptic locus. Cutting it with $c < g - 2$ divisors and pushing-down, we find relations in codimension c. There are many possible choices for the c divisors; e.g., one can take c diagonals, in which case the possibilities correspond to the partitions of c. It appears that these relations (together with the relation mentioned in (a) expressing κ_{g-2} in lower kappa's) generate the entire ideal of relations between the kappa's, but the computations are quite cumbersome.

Concluding remarks:

a. As we saw before, Looijenga's theorem (Theorem 1) in degree $g-1$ implies Diaz's upper bound $g-2$ for the dimension of a complete subvariety of $\mathcal{M}_g$. One may view Theorem 2 as indicating that there is no intersection-theoretical obstruction to the existence of a $(g-2)$-dimensional complete subvariety. Thinking along these lines, Theorem 1 in degree $g-2$ as well as Conjecture 1 put severe constraints on the possible $(g-2)$-dimensional complete subvarieties, whereas complete subvarieties of dimensions $\leq g/3$ are unconstrained from this point of view. In the known existence results, the genus is exponential in the dimension of the complete subvariety; for all $g \geq 3$, complete curves exist; $\mathcal{M}_8$ contains complete surfaces. Perhaps (for now) $\mathcal{M}_6$ and $\mathcal{M}_7$ are more natural places to look for complete surfaces than $\mathcal{M}_4$ and $\mathcal{M}_5$.

b. In degree $g-2$, we have found 'experimentally' several explicit proportionalities. Even deducing them from part (c) of Conjecture 1 appears to be non-trivial, so we are far from proving them.[8] We state the most relevant ones:

$$[\mathcal{H}_g]_Q = \frac{1}{2}[\mathcal{H}_g] = \frac{(2^{2g}-1)\,2^{g-2}}{(2g+1)(g+1)!}\,\kappa_{g-2}\,;$$

$$\lambda_{g-2} = |B_{2g-2}|\,\frac{(2g-1)\,2^{g-1}}{(2g-2)(g-1)!}\,\kappa_{g-2}\,.$$

Here $\mathcal{H}_g$ is the hyperelliptic locus. The second formula, together with Theorem 2, leads to

$$\lambda_{g-1}^3 = \frac{|B_{2g-2}B_{2g}|}{(2g-2)(2g)}\frac{1}{(2g-2)!}\,.$$

This is the contribution from the constant maps, as it occurs in the theory of counting curves of higher genus on threefolds (see [3], §5.13, (5.54)). Before, this number was known only for $g \leq 4$ (cf. [11]) and no conjectural formula was known.

Acknowledgements

I would like to thank Robbert Dijkgraaf, Bill Fulton, Gerard van der Geer, Ezra Getzler, Eduard Looijenga, Shigeyuki Morita, Ragni Piene, Piotr Pragacz, Michael Thaddeus, Chris Zaal and Don Zagier. This research has been made possible by a fellowship of the Royal Netherlands Academy of Arts and Sciences

---

[8] In fact, these formulas are proved now. For the proof of the latter two, see [14]. The hyperelliptic formula follows via (1) from a calculation similar to the one in [14], §3.2. This also gives a geometric proof of the non-vanishing of $R^{g-2}(\mathcal{M}_g)$.

and was carried out at the Universiteit van Amsterdam. The first version of this paper was completed at the Institut Mittag-Leffler, Djursholm, Sweden.

Bibliography

[1] E. Arbarello, M. Cornalba, P.A. Griffiths and J. Harris, *Geometry of Algebraic Curves I*, Grundlehren der math. Wiss., Volume 267, Springer, Berlin 1985.

[2] E. Arbarello, *Weierstrass points and moduli of curves*, Compos. Math. 29 (1974), 325–342.

[3] M. Bershadsky, S. Cecotti, H. Ooguri and C. Vafa, *Kodaira-Spencer Theory of Gravity and Exact Results for Quantum String Amplitudes*, Commun. Math. Phys. 165 (1994), 311–428.

[4] S. Diaz, *Exceptional Weierstrass points and the divisor on moduli space that they define*, Mem.Amer.Math.Soc. 327, 1985.

[5] S. Diaz, *A bound on the dimensions of complete subvarieties of* $\mathcal{M}_g$, Duke Math. J. 51 (1984), 405–408.

[6] R. Dijkgraaf, *Some facts about tautological classes*, private communication (November 1993).

[7] D. Edidin, *The codimension-two homology of the moduli space of stable curves is algebraic*, Duke Math. J. 67 (1992), 241–272.

[8] T. Eguchi, K. Hori, C.-S. Xiong: *Quantum cohomology and Virasoro algebra*. Phys. Lett. B 402 (1997), 71–80.

[9] C. Faber, *Chow rings of moduli spaces of curves I: The Chow ring of* $\overline{\mathcal{M}}_3$, Ann. of Math. 132 (1990), 331–419.

[10] C. Faber, *Chow rings of moduli spaces of curves II: Some results on the Chow ring of* $\overline{\mathcal{M}}_4$, Ann. of Math. 132 (1990), 421–449.

[11] C. Faber, *Intersection-theoretical computations on* $\overline{\mathcal{M}}_g$, in *Parameter Spaces* (Editor P. Pragacz), 71–81, Banach Center Publications, Volume 36, Warszawa 1996.

[12] C. Faber, *A non-vanishing result for the tautological ring of* $\mathcal{M}_g$, preprint 1995, math.AG/9711219.

[13] C. Faber, *Algorithms for computing intersection numbers on moduli spaces of curves, with an application to the class of the locus of Jacobians*, in *New Trends in Algebraic Geometry* (eds. F. Catanese, K. Hulek, C. Peters, M. Reid), 29–45, Cambridge University Press, 1999

[14] C. Faber, R. Pandharipande, *Hodge integrals and Gromov-Witten theory*, preprint 1998, math.AG/9810173.

[15] C. Faber, R. Pandharipande. In preparation.

[16] C. Faber, D. Zagier, paper on the relations in the tautological ring. In preparation.

[17] W. Fulton, *Intersection Theory*, Ergebnisse der Math., Volume 2, Springer, Berlin 1984.

[18] E. Getzler, R. Pandharipande: *Virasoro constraints and the Chern classes of the Hodge bundle.* Nucl. Phys. B 530 (1998), 701–714.

[19] A. Grothendieck, *Standard conjectures on algebraic cycles*, in *Algebraic Geometry, Bombay 1968*, Oxford University Press, Oxford 1969.

[20] J. Harer, *The second homology group of the mapping class group of an orientable surface*, Invent. Math. 72 (1983), 221–239.

[21] J. Harer, *Improved stability for the homology of the mapping class groups of surfaces*, preprint, Duke University, 1993. (http://www.math.duke.edu/preprints/).

[22] J. Harris, D. Mumford, *On the Kodaira dimension of the moduli space of curves*, Invent. Math. 67 (1982), 23–86.

[23] J. Horne, *Intersection theory and two-dimensional gravity at genus 3 and 4*, Modern Phys. Lett. A 5 (1990), 2127–2134.

[24] E. Izadi, *The Chow ring of the moduli space of curves of genus 5*, in *The Moduli Space of Curves* (Editors R. Dijkgraaf, C. Faber, G. van der Geer), 267–304, Progress in Math., Volume 129, Birkhäuser, Boston 1995.

[25] U. Jannsen, *Motivic Sheaves and Filtrations on Chow Groups*, in *Motives* (Editors U. Jannsen, S. Kleiman, J.-P. Serre), 245–302, Proc. Symp. Pure Math., Volume 55, Part 1, A.M.S., Providence 1994.

[26] M. Kontsevich, *Intersection theory on the moduli space of curves and the matrix Airy function*, Comm. Math. Phys. 147 (1992), 1–23.

[27] E. Looijenga, *On the tautological ring of $\mathcal{M}_g$*, Invent. Math. 121 (1995), 411–419.

[28] S. Morita, *Structure of the mapping class groups of surfaces: a survey and a prospect*, preprint 1999.

[29] D. Mumford, *Towards an enumerative geometry of the moduli space of curves*, in *Arithmetic and Geometry* (Editors M. Artin and J. Tate), 271–328, Part II, Progress in Math., Volume 36, Birkhäuser, Basel 1983.

[30] M. Pikaart, *An orbifold partition of $\overline{\mathcal{M}}_g^n$*, in *The Moduli Space of Curves* (Editors R. Dijkgraaf, C. Faber, G. van der Geer), 467–482, Progress in Math., Volume 129, Birkhäuser, Boston 1995.

[31] Z. Ran, *Curvilinear enumerative geometry*, Acta Math. 155 (1985), 81–101.

[32] E. Witten, *Two dimensional gravity and intersection theory on moduli space*, Surveys in Diff. Geom. 1 (1991), 243–310.

Correspondences between Moduli Spaces of Curves

Eduard Looijenga

Abstract

The moduli space of (possibly ramified) covers of nonsingular complex projective curves of a fixed topological type defines a correspondence between moduli spaces of pointed curves. We study the action of such a correspondence on the cohomology of these moduli spaces, where we pay special attention to what happens in the stable range.

1 Introduction

In this paper we begin the study of correspondences between moduli spaces of curves. The idea is simple enough: there is a moduli stack parameterizing unramified covers $\tilde{C} \to C$ of smooth (say, complex) projective curves of a fixed topological type. If the genera are $\tilde{g}$ and g respectively, then this moduli space defines a one-to-finite correspondence from $\mathcal{M}_g$ to $\mathcal{M}_{\tilde{g}}$. The case $g = 1$ is at the same time special and classical: then $\tilde{g} = 1$ also and we are dealing with Hecke correspondences. They generate an algebra. But if $g \geq 2$, then $\tilde{g} > g$, and the correspondences no longer form an algebra, but only an additive category (with an object for every genus $g \geq 2$). Yet, by virtue of Harer's stability theorem, which states that $H^k(\mathcal{M}_g)$ is independent of g when g is large compared to k, such correspondences will act on the cohomology of $\mathcal{M}_g$ in the stable range. It is not difficult to compute this action on a monomial in the so-called Miller-Morita-Mumford classes; we find that they are in fact common eigenvectors for these correspondences (Proposition 3.1). The precise result is best stated in terms of the Hopf algebra structure on the stable cohomology: the Miller-Morita-Mumford classes generate a Hopf subalgebra, the *tautological subalgebra*, and a correspondence defined by a degree d cover acts in the stable range of this tautological subalgebra as the Hopf algebra automorphism ψ_d which on the primitive part is multiplication by d.

Now Mumford has conjectured that the stable cohomology algebra is no bigger than the tautological subalgebra. As his conjecture is still open, one may try to show that correspondences act in the same way on the stable cohomology as they do on the tautological part. (And if one does not believe in the Mumford conjecture, one may think of Proposition 3.1 as a possible means to distinguish Miller-Morita-Mumford classes from other primitive classes.) At any rate, we prove that such a result is true for the simplest covers, namely the Galois covers of prime degree p, and then only in a suitable p-adic sense. But our motivation has another, though related source as well: The images of these correspondences define interesting algebraic cycles of high codimension on moduli spaces of stable curves that apparently have not been looked at before. Our ultimate goal is to understand these cycles and to provide a motivic context (both in statement and proof) for the stability theorems. At present, this is still a dream.

In the present paper we limit ourselves to characteristic zero. But as the case of genus one suggests, it may be expected that in positive characteristic we have new features, especially regarding the interplay between Hecke correspondences and the Frobenius.

Another observation we left unpursued is that the correspondences considered here have analogues for the Kontsevich moduli spaces that support the Gromov-Witten invariants.

Let us now briefly review the contents of the sections. In Section 2 we set up the Hecke category framework associated with unramified covers of smooth complex projective curves. The following section describes the action of the correspondences on the tautological subalgebra. In Section 4 we associate to a finite abelian group A a correspondence in every genus and we show that if we do this for A cyclic of prime order p, then the action of these correspondences on the stable cohomology converges in a p-adic sense to ψ_p if the genus of the base curve tends to ∞. We also consider for a positive integer d the abelian cover of degree d^{2g} of C given by $H_1(C; \mathbb{Z}/d)$. This defines a correspondence which is in fact a morphism. Its composite with $\psi_{d^{2g}}^{-1}$ stabilizes and stably this gives a Hopf algebra endomorphism. The proof is postponed to Section 5, where we state and prove more general results involving correspondences associated to possibly nonabelian ramified covers. (For optimal use of the stability theorem it is best to allow the covers to ramify.) We here also point out that these correspondences naturally act on the local systems defined by conformal blocks.

The final Section 6 raises what we believe is an interesting question: according to Kontsevich the cohomology of the disjoint union of moduli spaces $\mathcal{S}_n \backslash \mathcal{M}_g^n$ can be identified with the stable primitive cohomology of a graded Lie algebra pair; certain correspondences therefore act on the latter in a way that shift the grade and the dimension by the same amount. The problem is to understand this action in terms of the Lie algebra pair.

It is appropriate to mention here work of Biswas-Nag-Sullivan. Although their paper [2] differs in spirit and goals from ours, there is a relationship all the same. They investigate among other things identities (e.g., a form of the Mumford isomorphism) on curves that are invariant under finite unramified covers.

Acknowledgements. This paper has its origin in some correspondence with Carel Faber around 1993, during which Proposition 3.1 was established. My interest in the matter was revived by some email exchanges with Dick Hain in Spring 1997.

Most of this work was done while I was visiting and supported by the Institut des Hautes Études Scientifiques in Bures-sur-Yvette (France). I am very grateful to this Institute and its staff for providing such marvelous working conditions and I thank my home University for granting me a sabbatical leave during that period.

Finally, I thank Carel Faber for his comments on the penultimate e-script version of this paper.

Notation: If X is a space or a group, then $H_{\bullet}(X)$ denotes its rational homology and $H_{\bullet}^{\mathbb{Z}}(X)$ resp. $H_{\mathbb{Z}}^{\bullet}(X)$ its integral homology resp. cohomology modulo torsion. Similarly, if X is variety, then $\mathrm{CH}_k(X)$ stands for the kth Chow group for rational equivalence, tensorized with $\mathbb{Q}$. If X is smooth of pure dimension d, then $\mathrm{CH}^k(X) := \mathrm{CH}_{d-k}(X)$, a definition that extends in an evident manner to the case where X is the quotient of a smooth variety by a finite group action.

It is always understood that the surfaces under consideration are oriented, and that the maps between them respect the orientations.

2   Correspondences for Mapping Class Groups: Closed Surfaces

2.1   The mapping class category for closed surfaces

For any integer $g \geq 2$, we fix a connected oriented closed differentiable surface S_g of genus $g \geq 2$ with base point $* \in S_g$. We write π_g for its fundamental group. The orientation of S_g defines a natural generator of $H_2(\pi_g; \mathbb{Z})$, referred to as the *orientation of* π_g. The mapping class group $\Gamma(S_g, *)$ of $(S_g, *)$ can be identified with the group $\mathrm{Aut}^+(\pi_g)$ of orientation preserving automorphisms of π_g and the mapping class group $\Gamma(S_g)$ of S_g with the quotient $\mathrm{Out}^+(\pi_g)$ of $\mathrm{Aut}^+(\pi_g)$ by the subgroup of inner automorphisms $\mathrm{Int}(\pi_g)$. We define the *mapping class category* Γ as follows: its objects are the positive integers and for two such integers k, l, the morphism set $\Gamma(k, l)$ is the set of orientation preserving monomorphisms $\pi_{l+1} \to \pi_{k+1}$ modulo inner automorphisms of π_{k+1}. Composition is defined in the obvious way. The theory of covering maps shows that to give an orientation preserving monomorphism $\pi_{l+1} \to \pi_{k+1}$ up to inner automorphism amounts to giving an orientation preserving covering map $S_{l+1} \to S_{k+1}$ up to isotopy. The Hurwitz formula tells us that its degree is equal to l/k. So if $\mathbb{N}^\times$ denotes the semigroup of positive integers under multiplication, then we have a natural forgetful functor

$\Gamma \to \mathbb{N}^{\times}$. Notice that by taking $k = l$, we see that $\Gamma(k,k) = \text{Aut}_{\Gamma}(k)$ can be identified with $\Gamma(S_{k+1})$.

2.2 The category of mapping class correspondences

Assigning to k the cohomology of the mapping class group $\text{Aut}_{\Gamma}(k)$ does not define a functor from Γ to the category of abelian groups. In order that this be the case, we must replace the composition law of Γ by a convolution, and this leads to the notion of a Hecke category which we presently define.

For $f \in \Gamma(k,l)$, let $\tilde{f} : \pi_{l+1} \to \pi_{k+1}$ be a representative and $F : (S_{l+1},*) \to (S_{k+1},*)$ a corresponding covering. If $F' : (S_{l+1},*) \to (S_{k+1},*)$ is another covering, then we say that F' is *equivalent* to F if there exist sense preserving diffeomorphisms of source and target which carry one onto the other. So the equivalence class of F is described by the double coset $\Gamma(S_{k+1},*)\tilde{f}\Gamma(S_{l+1},*)$, or what amounts to the same, by $D_f := \Gamma(l,l)f\Gamma(k,k)$.

Let us write $\text{Aut}_{\Gamma}(f)$ for the group of $(u,\tilde{u}) \in \Gamma(k,k) \times \Gamma(l,l)$ satisfying $fu = \tilde{u}f$. It can be identified with the connected component group of the group of pairs $(U,\tilde{U})$ with U a sense preserving diffeomorphism of S_{k+1} and $\tilde{U}$ a sense preserving diffeomorphism of S_{l+1} which is a lift over F of U. If U is the identity, then $\tilde{U}$ must be a covering transformation and vice versa. In particular, the kernel of the projection $p : \text{Aut}_{\Gamma}(f) \to \Gamma(k,k)$ is finite. Since nontrivial covering transformations define nontrivial mapping classes, the other projection $q : \text{Aut}_{\Gamma}(f) \to \Gamma(l,l)$ is injective.

Lemma 2.1. *The image of* $\text{Aut}_{\Gamma}(f)$ *in* $\Gamma(k,k)$ *is a subgroup of finite index. This index is equal to the number of* $\Gamma(l,l)$-*cosets in the double coset* $D_f = \Gamma(l,l)f\Gamma(k,k)$.

Proof. Let $d = l/k$ be the degree of f. Then the image of $\tilde{f}$ is a subgroup of π_{k+1} of index d. The collection $\tilde{C}$ of subgroups of π_{k+1} of index d is finite. It is acted on by the mapping class group $\text{Aut}^{+}(\pi_{k+1})$, and the orbit set $C := \text{Int}(\pi_{k+1})\backslash\tilde{C}$ is acted on by $\text{Out}^{+}(\pi_{k+1}) = \Gamma(k,k)$. The image of $\tilde{f}$ defines an element $[f] \in C$ and it is clear that $u \in \Gamma(k,k)$ extends to an automorphism of f if and only if it fixes $[f]$. Moreover, the assignment $\psi \in D_f \mapsto [\psi] \in C$ induces a bijection from $\Gamma(l,l)f\Gamma(k,k)/\Gamma(l,l)f$ onto the $\Gamma(k,k)$-orbit of $[f]$. Both assertions follow. $\square$

We call the integer appearing in this lemma the *mass* of D_f and denote it by $\mu(D_f)$. Notice that this mass is one if and only if the image of $\tilde{f}$ is $\text{Aut}^{+}(\pi_{k+1})$-invariant. This is why we then say that f (or D_f) is *invariant*.

Example 2.2. We consider the case where f is represented by a monomorphism $\pi_{l+1} \to \pi_{k+1}$ whose image is a normal subgroup with cyclic quotient of order $d = l/k$. So f defines an order d subgroup $C(f) \subset H^1(S_{k+1}; \mathbb{Z}/d)$. This subgroup only depends on the coset $\Gamma(l,l)f$ and conversely, the subgroup determines the coset. If we let f run over the double coset $D_f := \Gamma(l,l)f\Gamma(k,k)$, then $C(f)$ runs over all such subgroups. So the mass of D_f is the number of order d subgroups of $(\mathbb{Z}/d)^{2k+2}$, i.e., $\phi_{2k+2}(d)/\phi_1(d)$, where $\phi_r(m)$ is a generalized Euler indicator: it is the number of elements in $(\mathbb{Z}/m)^r$ of exact order m:

$$\phi_r(m) = m^r \prod_{p|m, p \text{ prime}} (1 - p^{-r}). \tag{2.1}$$

The lemma implies that the composite of two left cosets is a finite union of left cosets: if $f \in \Gamma(k,l)$ and $\psi \in \Gamma(l,m)$ and $D_g = \cup_i \Gamma(m,m)g_i$, then

$$\Gamma(m,m)g\Gamma(l,l)f = \cup_i \Gamma(m,m)g_i f.$$

If the cosets $\Gamma(m,m)g_i$ are mutually disjoint, then so are the $\Gamma(m,m)g_i f$.

We introduce a "Hecke quotient" of Γ. Let $\mathcal{L}(k,l)$ denote the abelian group of $\mathbb{Z}$-valued functions on $\Gamma(k,l)$ spanned by the characteristic functions $E_{\Gamma(l,l)f}$ of the left cosets. We have a natural bilinear map $\mathcal{L}(l,m) \times \mathcal{L}(k,l) \to \mathcal{L}(k,m)$ defined by convolution: given $f \in \Gamma(k,l)$ and $\psi \in \Gamma(l,m)$, then define the image of $E_{\Gamma(m,m)\psi} \otimes E_{\Gamma(l,l)f}$ as the characteristic function of $\Gamma(m,m)\psi\Gamma(l,l)f$. This is in $\mathcal{L}(k,m)$ by the remark above. The composition is obviously associative and so we have an additive category $\mathcal{L}$. The homomorphism $\mathcal{L}(k,l) \to \mathbb{Z}$ which takes the value one on every generator $E_{\Gamma(l,l)f}$ will be called the *mass functional* and denoted by μ also.

The functor $\Gamma \to \mathbb{N}^\times$ extends to an additive functor from $\mathcal{L}$ to the free additive category $\mathbb{Z}[\mathbb{N}^\times]$ generated by $\mathbb{N}^\times$. The ring of additive endofunctors of $\mathbb{Z}[\mathbb{N}^\times]$ has as an additive basis the functors "multiplication by d", ψ_d (here d runs over the positive integers). This is clearly a polynomial algebra with generators the ψ_p with p prime. We may (and often will) think of this ring as a quotient category of $\mathbb{Z}[\mathbb{N}^\times]$ which identifies the unique morphism $k \to kd$ with ψ_d. We denote the composite functor $\mathcal{L} \to \mathbb{Z}[\psi_p : p \text{ prime}]$ by χ and call it the *Adams character*. So if $f \in \Gamma(k,l)$, then $\chi(E_{\Gamma(l,l)f}) = \psi_{l/k}$.

We introduced the category $\mathcal{L}$ only for an auxiliary purpose, for we will rather be concerned with a subcategory of it. This subcategory $\mathcal{H}$ shall have the same object set (i.e., the positive integers), but the morphism set $\mathcal{H}(k,l)$ is to be the submodule of $\mathcal{L}(k,l)$ spanned by the characteristic functions E_{D_f} of the double cosets D_f. Convolution preserves these submodules so that a subcategory is defined indeed. Notice that $\chi(E_{D_f}) = \mu(D_f)\psi_{l/k}$.

Another useful category is the subcategory of $\mathcal{L} \otimes \mathbb{Q}$ generated by the characteristic functions of double cosets divided by their mass:

$$T_{D_f} := \mu(D_f)^{-1}E_{D_f} \in \mathcal{L}(k,l) \otimes \mathbb{Q}. \tag{2.2}$$

It is clear that this category, which we denote by $\tilde{\mathcal{H}}$, contains $\mathcal{H}$. Notice that $\mu(D_f) = \mu(E_{D_f})$, so that the linear extension of the mass functional takes unit value on T_{D_f}. The Adams character naturally extends to $\tilde{\mathcal{H}}$ and maps T_{D_f} to $\psi_{l/k}$.

The Hecke category $\mathcal{H}$ has a universal property: suppose that we are given a covariant functor Y from Γ to the category of spaces. So $Y(k)$ is a space with $\Gamma(k,k)$-action and we may form its orbit space $\overline{Y}(k)$. Then every double coset $D = \Gamma(l,l)f\Gamma(k,k)$ defines a one-to-finite correspondence $E_D : \overline{Y}(k) \rightrightarrows \overline{Y}(l)$: it assigns to the orbit $\Gamma(k,k)x$ the finite union of $\Gamma(l,l)$-orbits $Y(\Gamma(l,l)f\Gamma(k,k))x = (\Gamma(l,l)Y(f_i)x)_i$, where $\Gamma(l,l)f_i$ runs over the distinct left cosets in D. Often such a one-to-finite correspondence defines a map on homology and so we get a covariant additive functor from $\mathcal{H}$ to the category of $\mathbb{Q}$-vector spaces.

The basic example is the Teichmüller functor $\mathcal{X}$ which assigns to k the Teichmüller space $\mathcal{X}_{k+1}$ of conformal structures on S_{k+1} modulo isotopies. An isotopy class of orientation preserving coverings $F : S_{l+1} \to S_{k+1}$ defines an analytic morphism $\mathcal{X}_{k+1} \to \mathcal{X}_{l+1}$ (defined by lifting the conformal structure). This in fact defines a functor from Γ to an analytic category. The mapping class group $\Gamma(k,k)$ acts properly discontinuously on $\mathcal{X}_{k+1}$ and the orbit space $\mathcal{M}_{k+1}$ can be regarded as the moduli space of nonsingular complex projective curves of genus $k + 1$. It is naturally a quasi-projective orbifold. A double coset $D = \Gamma(l,l)f\Gamma(k,k)$ defines a one-to-finite correspondence E_D from $\mathcal{M}_{k+1} \to \mathcal{M}_{l+1}$. This correspondence is finite algebraic; a more concrete description is

$$\mathcal{M}_{k+1} \xleftarrow{p} \mathcal{M}(D) \xrightarrow{q} \mathcal{M}_{l+1}. \tag{2.3}$$

Here $\mathcal{M}(D)$ is the moduli space of pointed unramified covers $\tilde{C} \to C$, with C a nonsingular complex projective connected curve, which on fundamental groups induce a map equivalent to f, and the projections are the obvious forgetful maps. Note that since the curve $\tilde{C}$ has in general a nontrivial group of automorphisms, q may map to the singular locus of $\mathcal{M}_{l+1}$.

Lemma 2.3. *The map $p : \mathcal{M}(D) \to \mathcal{M}_{k+1}$ is an unramified covering in the orbifold sense of degree $\mu(D)$ and $q : \mathcal{M}(D) \to \mathcal{M}_{l+1}$ is finite and generically injective. Assigning to D the correspondence qp^{-1} defines an additive functor from $\mathcal{H}$ to the category of correspondences.*

Proof. Let C be a nonsingular complex projective curve of genus $k + 1$. Then the fiber of p over $[C] \in \mathcal{M}_{k+1}$ parameterizes the unramified covers of C equivalent to $F : S_{l+1} \to S_{k+1}$. These are $\mu(D)$ in number by Lemma 2.1. A nonsingular complex-projective curve $\tilde{C}$ of genus $l + 1$ is in at most a finite number of ways realized as an unramified cover of a curve of genus $k + 1$ and if that number is positive, then generically it is one. $\square$

So E_D acts as $q_*p^!$. Since $p_*p^!$ is multiplication by $\mu(D)$, the map $\mu(D)^{-1}p^!$ realizes the homology of $\mathcal{M}_{k+1}$ as a direct summand of the homology of $\mathcal{M}(D)$. So from a motivic point of view it is perhaps better to replace E_D by T_D, which acts as $\mu(D)^{-1}q_*p^!$. We refer to T_D as the *normalized Hecke correspondence* defined by D. Notice that it respects the augmentations. The map T_D also acts on the Chow groups (after tensorizing them with $\mathbb{Q}$).

There is a natural extension of the diagram (2.3) to the Deligne-Mumford compactifications:

$$\overline{\mathcal{M}}_{k+1} \xleftarrow{\bar{p}} \overline{\mathcal{M}}(D) \xrightarrow{\bar{q}} \overline{\mathcal{M}}_{l+1}.$$

Here $\overline{\mathcal{M}}(D)$ is simply the normalization of $\mathcal{M}(D)$ over $\overline{\mathcal{M}}_{k+1}$. This variety parameterizes (perhaps noneffectively) *admissible covers* of stable curves (in the sense of [6], in particular, ramification is only allowed in singular points and locally given by $w^k = xy = 0$). This implies that q extends to a morphism $\overline{\mathcal{M}}(D) \to \overline{\mathcal{M}}_{l+1}$. The corresponding extension of E_D resp. T_D is denoted $\overline{E}_D$ resp. $\overline{T}_D$.

Question 2.4. Is the class in $\mathrm{CH}^{3(l-k)}(\overline{\mathcal{M}}_{l+1})$ defined by the image of $\overline{T}_D$ in the tautological algebra of $\overline{\mathcal{M}}_{l+1}$ (in the sense of Section 5 of [3])?

3 Action of the Correspondences on the Tautological Classes

Let $f : \mathcal{C} \to S$ be a family of stable complex curves. Denote by ω_f its relative dualizing sheaf and let $K(f) \in \mathrm{CH}^1(\mathcal{C})$ be its first Chern class. Then the nth Mumford class of this family is $\kappa(f)_n := f_*(K(f)^{n+1}) \in \mathrm{CH}^n(S)$. This definition extends to the orbifold setting and so we have also defined for $g \geq 2$, $\overline{K} \in \mathrm{CH}^1(\overline{\mathcal{M}}_g^1)$, $K \in \mathrm{CH}^1(\mathcal{M}_g^1)$ and a universal Mumford class $\overline{\kappa}_n \in \mathrm{CH}^n(\overline{\mathcal{M}}_g)$, $\kappa_n \in \mathrm{CH}^n(\mathcal{M}_g)$. (The genus is deliberately left out from the notation.)

Proposition 3.1. Let $f \in \Gamma(k,l)$. Then $\overline{T}_{D_f}$ takes the monomial $\overline{\kappa}_{n_1}\overline{\kappa}_{n_2} \cdots \overline{\kappa}_{n_r}$ to $(\frac{l}{k})^r \overline{\kappa}_{n_1}\overline{\kappa}_{n_2} \cdots \overline{\kappa}_{n_r}$.

It is easy to see that this proposition follows from the following assertion:

Proposition 3.2. *Suppose we are given a family of stable complex curves* $\pi : \mathcal{C} \to S$ *as above with smooth base, a finite surjective map* $q : \tilde{S} \to S$ *of degree* μ *and an unramified covering* $p : \tilde{\mathcal{C}} \to q^*\mathcal{C}$ *of degree* d:

$$
\begin{array}{ccccc}
\tilde{\mathcal{C}} & \xrightarrow{p} & q^*\mathcal{C} & \xrightarrow{\pi^*q} & \mathcal{C} \\
{\scriptstyle \tilde{\pi}}\downarrow & & {\scriptstyle q^*\pi}\downarrow & & {\scriptstyle \pi}\downarrow \\
\tilde{S} & = & \tilde{S} & \xrightarrow{q} & S,
\end{array}
$$

where $\tilde{\pi} = (q^*\pi)p$. *Then* $(\pi^*q)_* p_*$ *maps* $K(\tilde{\pi})^n$ *to* $\mu d K(\pi)^n$ *and* q_* *maps the monomial* $\kappa_{n_1}(\tilde{\pi})\kappa_{n_2}(\tilde{\pi}) \cdots \kappa_{n_r}(\tilde{\pi}))$ *to* $\mu d^r \kappa_{n_1}(\pi)\kappa_{n_2}(\pi) \cdots \kappa_{n_r}(\pi)$.

Proof. First observe that $\omega_{\tilde{\pi}}$ is the coherent pull-back of ω_π. So $K(\tilde{\pi})^n$ is the pull-back of $K(\pi)^n$. Since p_*p^* resp. q_*q^* is multiplication by d resp. μ, it follows that $(\pi^*q)_* p_*(K(\tilde{\pi})^n) = \mu d K(\pi)^n$. Applying π_* to this yields $q_* \overline{\kappa}_{n-1}(\tilde{\pi}) = \mu d \overline{\kappa}_{n-1}(\pi)$. In order to generalize this to any monomial in the kappa's, we consider the diagram of r-fold fibered products:

$$\tilde{C}^{(r)} \xrightarrow{p^{(r)}} q^*C^{(r)} \longrightarrow C^{(r)}$$

$$\downarrow \qquad\qquad \downarrow \qquad\qquad \downarrow$$

$$\tilde{S} \;=\!=\!= \; \tilde{S} \xrightarrow{q} S.$$

If $\pi_i : C^{(r)} \to C^{(r-1)}$ omits the ith factor, then we have defined $K(\pi_i) \in \mathrm{CH}^1(\mathcal{C}(r))$. Since $K(\pi_1)^{n_1+1}K(\pi_2)^{n_2+1} \cdots K(\pi_r)^{n_r+1} \in \mathrm{CH}(\mathcal{C}(r))$ pulls back to the corresponding class on $\tilde{C}^{(r)}$, and since $p^{(r)}$ has degree d^r, an argument as above shows that q_* has the asserted effect on $\kappa_{n_1}(\tilde{\pi})\kappa_{n_2}(\tilde{\pi}) \cdots \kappa_{n_r}(\tilde{\pi})$. $\qquad\square$

According to Harer ([4], [5]), the homology groups $H_s(\Gamma(S_g); \mathbb{Z})$ can for fixed s be canonically identified with each other when $s \le cg$ for some positive constant c (c about $\frac{2}{3}$ will do). The more precise result can be stated as follows. If S is a connected orientable compact surface possibly with boundary, then let $\Gamma(S)$ denote the mapping class group of S relative its boundary. An embedding $S \to S'$ induces a homomorphism of mapping class groups $\Gamma(S) \to \Gamma(S')$, and Harer shows in fact that this map induces an isomorphism on integral homology in degree $\le cg(S)$, where $g(S)$ denotes the genus of S. We denote the corresponding stable (co)homology groups mod torsion by $H_\bullet^{\mathbb{Z}}(\Gamma_\infty)$ resp. $H_{\mathbb{Z}}^\bullet(\Gamma_\infty)$. E. Miller observed that $H_\bullet^{\mathbb{Z}}(\Gamma_\infty)$ comes naturally with the structure of a Hopf algebra; the coproduct is the standard one (it comes from the diagonal embedding) and the product from embedding two surfaces S_1 and S_2 as above with nonempty boundary in a third, S, say. This yields a homomorphism $\Gamma(S_1) \times \Gamma(S_2) \to \Gamma(S)$ and thus produces in the stable range a map $H_r^{\mathbb{Z}}(\Gamma_\infty) \otimes H_s^{\mathbb{Z}}(\Gamma_\infty) \to H_{r+s}^{\mathbb{Z}}(\Gamma_\infty)$. Using Harer's theorem, it is easily seen to be independent of choices. It defines the (Hopf) product. As a Hopf algebra, $H_\bullet^{\mathbb{Z}}(\Gamma_\infty)$ is graded-bicommutative. Since $H_{\mathbb{Z}}^\bullet(\Gamma_\infty)$ is the graded dual of $H_\bullet^{\mathbb{Z}}(\Gamma_\infty)$, it is also a Hopf algebra. Miller and Morita have shown that the cohomological Mumford class κ_n stabilizes and defines a nonzero primitive integral element of $H_{\mathbb{Z}}^{2n}(\Gamma_\infty)$ (which is also denoted by κ_n): The coproduct sends κ_n to $\kappa_n \otimes 1 + 1 \otimes \kappa_n$. So these elements generate a Hopf subalgebra $H_{\mathrm{taut}}^\bullet(\Gamma_\infty; \mathbb{Z})$ of $H_{\mathbb{Z}}^\bullet(\Gamma_\infty)$, called the *tautological algebra*. Mumford has conjectured that the two coincide after we tensorize with $\mathbb{Q}$.

As any commutative Hopf algebra, $H_\bullet^{\mathbb{Z}}(\Gamma_\infty)$ comes with "Adams operations": for a positive integer n, the nth Adams operation ψ_n is the composite

$$\psi_n : H_\bullet^{\mathbb{Z}}(\Gamma_\infty) \to H_\bullet^{\mathbb{Z}}(\Gamma_\infty)^{\otimes n} \to H_\bullet^{\mathbb{Z}}(\Gamma_\infty),$$

where the first map is $(n-1)$-fold iteration of the coproduct and the second is multiplication. This map is simply the Hopf algebra endomorphism which on the primitive part is multiplication by n. So ψ_n is invertible on $H_\bullet(\Gamma_\infty)$ and $\psi_n\psi_m = \psi_{nm}$. Thus we have defined a ring homomorphism, the *Adams action*,

$$\mathbb{Z}[\psi_p : p \text{ prime}] \to \mathrm{End}(H_\bullet^{\mathbb{Z}}(\Gamma_\infty)). \tag{3.1}$$

For any double coset $D = \Gamma(l,l)f\Gamma(k,k)$, E_D induces an endomorphism of H_s if s is in the stable range with respect to $k+1$. Then Proposition 3.1 amounts to the statement that in the stable range T_D acts as $\psi_{\deg(D)}^*$ on the tautological classes. So we can reformulate 3.1 as follows:

Proposition 3.3. *In the stable range, the Hecke category $\tilde{\mathcal{H}}$ preserves the tautological algebra and the action factors through the Adams character: it is given by composing χ with the Adams action.*

4 Hecke Operators Attached to Finite Abelian Covers

As long as the Mumford conjecture is open, it is worthwhile to attempt to prove that the above proposition holds on all of $H^\bullet(\Gamma_\infty)$. We have not succeeded in this, but we will show that in certain cases a weak form of this property holds.

Fix a finite abelian group A. Given a positive integer k, then every element u of $H^1(S_{k+1}; A)$ can be thought of as a homomorphism $\pi_{k+1} \to A$ and thus defines an abelian covering of S_{k+1}. The degree d_u of this covering is of course the order of the image of the map $\pi_{k+1} \to A$. We thus get a well-defined double coset $D(u)$ and hence an element $E_{D(u)} \in \mathcal{H}(k, d_u k)$ and its normalization $T_{D(u)} \in \tilde{\mathcal{H}}(k, d_u k)$. We now let s be in the stable range with respect to k and consider the expression

$$\mathcal{T}_{A,k} := |A|^{-2k-2} \sum_{u \in H^1(S_{k+1}, A)} \psi_{|A|/d_u} T_{D(u)}, \tag{4.1}$$

viewed as an endomorphism of $H_s(\Gamma_\infty)$.

Proposition 4.1. *The element $\mathcal{T}_{A,k}$ is independent of k and the resulting endomorphism $\mathcal{T}_A$ of $H_\bullet(\Gamma_\infty)$ is in fact an algebra endomorphism.*

Notice that we do not claim that $\mathcal{T}_A$ preserves the coproduct. We postpone the proof of 4.1 to 5.4, where we will in fact prove a nonabelian generalization of this result. The case that interests us most is when A is cyclic of order d. We then write $T_{d,k}$ for the associated normalized operator. (It follows from 2.2 that the unnormalized operator $E_{d,k}$ is $\phi_{2k+2}(d)/\phi(d)T_{d,k}$, where ϕ_r is the Euler indicator defined in (2.1) that generalizes the usual one $\phi = \phi_1$.) So if we denote the linear combination (4.1) by $\mathcal{T}_d$, then

$$\mathcal{T}_d = d^{-2k-2} \sum_{m|d} \phi_{2k+2}(m)\psi_{d/m} T_{m,k}. \tag{4.2}$$

For instance, when $d = p$ is prime, then $\mathcal{T}_{p,k} = p^{-2k-2}((p^{2k+2} - 1)T_{p,k} + \psi_p)$. By means of Möbius inversion we can express $T_{d,k}$ as a weighted average of operators $\psi_* \mathcal{T}_*$: if $\mathcal{P}(d)$ is the set of primes dividing d, then

$$T_{d,k} \prod_{p \in \mathcal{P}(d)} (p^{2k+2} - 1) = \sum_{I \subset \mathcal{P}(d)} (-1)^{|I|} \operatorname{pr}(\mathcal{P}(d) - I)^{2k+2} \psi_{\operatorname{pr}(I)} \mathcal{T}_{d/\operatorname{pr}(I)}. \tag{4.3}$$

Here $\operatorname{pr}(I)$ stands for the product of the members of I (the empty product is 1 by convention). This formula gives $T_{d,k}$ a sense on all of $H_\bullet(\Gamma_\infty)$.

Corollary 4.2. *Let p be a prime and let d be a positive integer not divisible by p.*

(i) For every positive integer n the action of $T_{p^n,k}$ on $H_\bullet(\Gamma_\infty)$ converges to $\psi_p T_{p^{n-1}}$ in the p-adic topology as $k \to \infty$.

(ii) The action of $T_{pd,k} - \psi_p T_{d,k}$ on $H_\bullet(\Gamma_\infty)$ converges to zero in the p-adic topology as $k \to \infty$.

Proof. Formula (4.3) gives:

$$(p^{2k+2} - 1)T_{p^n,k} = p^{2k+2} T_{p^n} - \psi_p T_{p^{n-1}}.$$

Taking the limit for $k \to \infty$ yields the first assertion. For the second statement we note that

$$T_{pd,k}(p^{2k+2} - 1) \prod_{l \in \mathcal{P}(d)} (l^{2k+2} - 1) =$$
$$\sum_{I \subset \mathcal{P}(dp)} (-1)^{|I|} \operatorname{pr}(\mathcal{P}(dp) - I)^{2k+2} \psi_{\operatorname{pr}(I)} T_{pd/\operatorname{pr}(I)}.$$

Reducing this identity modulo p^{2k+2} reduces the sum on the righthand side to a sum over the subsets I of $\mathcal{P}(dp)$ containing p. Writing this as a sum over the subsets of J of $\mathcal{P}(d)$, we get

$$\sum_{J \subset \mathcal{P}(d)} -(-1)^{|J|} \operatorname{pr}(\mathcal{P}(d) - J)^{2k+2} \psi_p \psi_{\operatorname{pr}(J)} T_{d/\operatorname{pr}(J)}.$$

We recognize this expression as $-\psi_p T_{d,k} \prod_{l \in \mathcal{P}(d)}(l^{2k+2} - 1)$ and so we find that

$$T_{dp,k} \equiv \psi_p T_{d,k} \quad \mathrm{mod}\ p^{2k+2}.$$

The statement follows. $\qquad\qquad\qquad\qquad\qquad\qquad\qquad\qquad\qquad\qquad$ □

By Corollary 4.2, $T_{p,k} : H_r(\mathcal{M}_{k+1}) \to H_r(\mathcal{M}_{pk+1})$ is an isomorphism when k is large enough. Fix a positive integer d not divisible by the prime p. Then the direct limit of the system

$$\Sigma_p(d) : H_r(\mathcal{M}_{d+1}) \xrightarrow{T_{p,d}} H_r(\mathcal{M}_{pd+1}) \xrightarrow{T_{p,pd}} \cdots \longrightarrow H_r(\mathcal{M}_{p^s d+1}) \xrightarrow{T_{p,p^s d}} \cdots$$

is isomorphic to $H_r(\Gamma_\infty)$, but the isomorphism is not canonical (it depends on the choice of a sufficiently large integer of the form dp^s). However, there is one if we tensorize with the p-adic numbers:

Theorem 4.3. *The direct limit of the system $\Sigma_p(d) \otimes \mathbb{Q}_p$:*

$$H_r(\mathcal{M}_{d+1}; \mathbb{Q}_p) \xrightarrow{T_{p,d}} H_r(\mathcal{M}_{pd+1}; \mathbb{Q}_p) \xrightarrow{T_{p,pd}} \cdots \longrightarrow H_r(\mathcal{M}_{p^s d+1}; \mathbb{Q}_p) \xrightarrow{T_{p,p^s d}} \cdots$$

is canonically isomorphic to the stable cohomology $H_r(\Gamma_\infty; \mathbb{Q}_p)$.

Proof. For k in the stable range with respect to r, we define $u_k : H_r(\Gamma_\infty) \to H_r(\mathcal{M}_{k+1})$ as the composite of ψ_k with the natural isomorphism. Every such u_k is an isomorphism. If we identify the terms of large index of $\Sigma_p(d)$ with some $H_r(\mathcal{M}_{p^s d+1})$, then it is clear that $(u_{p^{t+1}d} - T_{p,p^t d} u_{p^t d})_t$ converges to zero in the p-adic topology. Hence $(u_{p^t d})_t$ induces an isomorphism of $H_r(\Gamma_\infty; \mathbb{Q}_p)$ onto the direct limit of $\Sigma_p(d) \otimes \mathbb{Q}_p$. □

A weakness of Theorem 4.3 is that it does not relate the isomorphisms for different choices of d. For instance, one would like to say that for a prime $\ell \neq p$, the Hecke correspondences $T_{\ell,*}$ map the system $\Sigma_p(d)$ to $\Sigma_{d\ell}(p)$, or at least in the p-adic limit. In other words, we would like the correspondences $T_{p,*}$ and $T_{\ell,*}$ to commute in a weak sense. Related to this is the question of whether the endomorphisms T_n commute for various n.

There is another series of Hecke correspondences of interest. For a positive integer d, the natural map $\pi_{k+1} \to H_1(S_{k+1}; \mathbb{Z}/d)$ defines an invariant covering of S_{k+1} of degree $d^{2(k+1)}$. The corresponding double coset is a left coset and so the associated Hecke correspondence $T_k(d)$ is in fact a morphism from $\mathcal{M}_{k+1}$ to $\mathcal{M}_{kd^{2(k+1)}+1}$.

Proposition 4.4. *The action of $\psi_{d^{2k+2}}^{-1} T_k(d)$ on $H_\bullet(\mathcal{M}_{k+1})$ in the stable range is independent of k. The resulting endomorphism $S(d)$ of $H_\bullet(\Gamma_\infty)$ is in fact a Hopf algebra endomorphism.*

The proof will be given in section 5 as a special case of a more general statement. The endomorphism $S(d)$ acts as the identity on the tautological subalgebra, but I do not know whether it is in fact the identity. I do not even know whether the $S(d)$'s mutually commute for varying d.

5   Correspondences via Ramified Covers

There is a much bigger category of correspondences acting on the stable cohomology of the mapping class groups if we allow the coverings to ramify. This makes fuller use of the stability theorem, since we will now also deal with the algebro-geometric analogues of surfaces with boundary. For this, we choose in $S_g - \{*\}$ a sequence $x_1, x_2, x_3, \ldots$ of distinct points and for every x_i a sense preserving linear isomorphism $v_i : T_{x_i} S_g \to \mathbb{C}$. The real oriented blow-up of S_g in $x_1, \ldots, x_n$ gives a surface $S_{g,n}$ with boundary and the mapping class group $\Gamma(S_{g,n})$ can be identified with the group $\Gamma(S_g; v_1, \ldots, v_n)$ of connected components of the group of sense preserving diffeomorphisms of S_g which fix v_i for $i = 1, \ldots, n$. The relevance of this remark is that to $(S_g; v_1, \ldots, v_n)$ there is naturally associated a moduli space: let $\mathcal{X}_{g,n}$ be the Teichmüller space of conformal structures on S_g extending $v_1, \ldots, v_n$, up to isotopy relative $(v_1, \ldots, v_n)$. This is a contractible complex manifold of complex dimension $3g - 3 + 2n$ and it is clearly acted on by the mapping class group $\Gamma(S_g; x_1, \ldots, x_n)$ resp. $\Gamma(S_g; v_1, \ldots, v_n)$. The natural map $\mathcal{X}_{g,n} \to \mathcal{X}_g$ is an equivariant analytic submersion. In between we have the Teichmüller space $\mathcal{X}_g^n$ of conformal structures on S up to isotopy relative $(x_1, \ldots, x_n)$; it is a contractible complex manifold of complex dimension $3g - 3 + n$. The action of the relevant mapping class group is properly discontinuous so that the orbit space $\mathcal{M}_g^n$ of $\mathcal{X}_g^n$ resp. $\mathcal{M}_{g,n}$ of $\mathcal{X}_{g,n}$ is an orbifold. This orbit space is the moduli space of tuples $(C; x_1, \ldots, x_n)$ resp. $(C; t_1, \ldots, t_n)$, where C is a nonsingular complex projective curve of genus g, and $x_1, \ldots, x_n$ are distinct points of C resp. $t_1, \ldots, t_n$ are nonzero tangent vectors at distinct points of C. (This notation may lead to confusion, since our $\mathcal{M}_g^n$ is what many algebraic geometers denote by $\mathcal{M}_{g,n}$, but it is in agreement with Harer's convention, which presently suits me better.). This moduli space interpretation brings us in the category of quasi-projective orbifolds: the maps

$$\mathcal{M}_{g,n} \to \mathcal{M}_g^n \to \mathcal{M}_g \tag{5.1}$$

are quasi-projective, and the first map is naturally a principal $(\mathbb{C}^\times)^n$-bundle in the orbifold sense.

Notice that $\mathcal{M}_g^n$ resp. $\mathcal{M}_{g,n}$ is a virtual classifying space for $\Gamma(S_g; x_1, \ldots, x_n)$ resp. $\Gamma(S_g; v_1, \ldots, v_n)$ and that (5.1) gives

$$\Gamma(S_{g,n}) \cong \Gamma(S_g; v_1, \ldots, v_n) \to \Gamma(S_g; x_1, \ldots, x_n) \to \Gamma(S_g)$$

on orbifold fundamental groups. Following Harer, the composite map induces an isomorphism on integral homology in the stable range. We conclude:

Proposition 5.1. *The projection $\mathcal{M}_{g,n} \to \mathcal{M}_g$ induces an isomorphism on rational cohomology in the stable range.*

An n-tuple of positive integers $d = (d_1, \ldots, d_n)$ determines a subgroup $\mu_d := \mu_{d_1} \times \cdots \times \mu_{d_n}$ of $(\mathbb{C}^\times)^n$. The orbit space $\mu_d \backslash \mathcal{M}_{g,n}$ can be interpreted as the moduli space of tuples $(C; u_1, \ldots, u_n)$ with C a nonsingular complex projective curve of genus g and u_i is a nonzero tangent vector in $T^{\otimes d_i} C$ such that the base points of $u_1, \ldots, u_n$ are distinct. It is a virtual classifying space whose orbifold fundamental group $\Gamma(S_g; v^{d_1}, \ldots, v^{d_n})$ can be described as the connected component group of the group of sense preserving diffeomorphisms of S_g which fix $x_1, \ldots, x_n$ and the maps $v_1^{d_1}, \ldots, v_n^{d_n}$.

The permutation group $\mathcal{S}_n$ acts on $\mathcal{M}_{g,n}$ by re-indexing the tangent vectors. This action is trivial on $\mathcal{M}_g$ and so the induced map $\mathcal{S}_n \backslash \mathcal{M}_{g,n} \to \mathcal{M}_g$ also induces an isomorphism on rational cohomology in the stable range.

5.1 Correspondences involving ramification

Let $f : S_{h,m} \to S_{g,n}$ be a finite, unramified, sense preserving covering. This covering determines a ramified covering $S_h \to S_g$, which we also denote by f.

We require compatibility with respect to our decorations v_i: if f ramifies in x_j, $1 \leq j \leq m$, of order d_j, and sends it to x_i, $1 \leq i \leq n$, then we want the pull-back of v_i under the d_j-jet of f at x_j to be equal to $v_j^{d_j}$. We remark that for $n > 0$ such a covering can be described as a functor between certain subgroupoids of the fundamental groupoids of $S_{h,m}$ and $S_{g,n}$ and that this can be used to define a mapping class category as we did in the undecorated case. Since we can do without that, we will not elaborate.

Let us define a *decorated* curve as a triple (C,X,v), where C is a complex projective curve, X a finite subset of its smooth part, and v a trivialization of the tangent bundle of C restricted to X. If we take C to be connected smooth of genus g and fix the cardinality n of X, then the moduli space of such curves can be identified with $S_n \backslash \mathcal{M}_{g,n}$. Allowing (C,X) to be stable as a pointed curve yields $S_n \backslash \overline{\mathcal{M}}_{g,n}$.

A *cover* $(\tilde{C}, \tilde{X}, \tilde{v}) \to (C,X,v)$ of two nonsingular decorated curves is a morphism $\pi : \tilde{C} \to C$ such that

(i) $\pi^{-1}(X) = \tilde{X}$,

(ii) X contains the discriminant of π,

(iii) if π ramifies of order d at $\tilde{p} \in \tilde{X}$, then v and $\tilde{v}^{\otimes d}$ define the same trivialization of $T_{\tilde{p}} \tilde{C}^{\otimes d}$.

It should be clear what we mean when we say that such a cover is equivalent to f. Let $\mathcal{M}(f)$ denote the moduli space of covers between nonsingular decorated curves which are equivalent to f. We then have forgetful morphisms

$$S_n \backslash \mathcal{M}_{g,n} \xleftarrow{p} \mathcal{M}(f) \xrightarrow{q} S_m \backslash \mathcal{M}_{h,m}$$

As in the undecorated case, p is a finite covering. In particular, qp^{-1} is a one-to-finite correspondence. We call the degree of p the *mass* of f and denote it by $\mu(f)$. Similarly we define the $\mathbb{Q}$-correspondence

$$T(f) := \mu(f)^{-1} qp^{-1} : S_n \backslash \mathcal{M}_{g,n} \rightrightarrows S_m \backslash \mathcal{M}_{h,m}$$

It acts on homology as $T(f)_* = \mu(f)^{-1} q_* p^!$. In view of the discussion at the beginning of this section, this defines an action on $H_\bullet(\Gamma_\infty)$ in the stable range.

The normalization $\overline{p} : \overline{\mathcal{M}}(f) \to S_n \backslash \overline{\mathcal{M}}_{g,n}$ of $\mathcal{M}(f)$ over $S_n \backslash \overline{\mathcal{M}}_{g,n}$ can still be interpreted as a moduli space (namely as one of admissible covers between stable decorated curves), and this shows that q extends to a morphism $\overline{q} : \overline{\mathcal{M}}(f) \to S_m \backslash \overline{\mathcal{M}}_{h,m}$. The latter is finite and so $T(f)$ extends as a $\mathbb{Q}$-correspondence $\overline{T}(f) : S_n \backslash \overline{\mathcal{M}}_{g,n} \rightrightarrows S_m \backslash \overline{\mathcal{M}}_{h,m}$.

Remark 5.2. Our correspondences lift to the local systems defined by conformal blocks. To explain, let us begin with going quickly through the relevant definitions (for details we refer to [1] and [8] and the references cited therein). First of all, we need a connected nonsingular complex projective curve C, a finite subset $X \subset C$ so that $U := C - X$ is affine and a semisimple complex Lie algebra $\mathfrak{g}$. Put $\mathfrak{g}(U) := \mathfrak{g} \otimes \mathcal{O}(U)$ and let $\mathfrak{g}_X$ be the completion of $\mathfrak{g}(U)$ along X, in other words, $\mathfrak{g}_X = \prod_{x \in X} \mathfrak{g}_x$, where $\mathfrak{g}_x$ is $\mathfrak{g}$ tensorized with the quotient field of the completed local ring of (C,x). This Lie algebra has a natural central extension $\hat{\mathfrak{g}}_X$ by $\mathbb{C}$, defined by the cocycle

$$(\sum_{x \in X} Y_x \otimes f_x, \sum_{x \in X} Z_x \otimes g_x) \mapsto \sum_{x \in X} \langle Y_x, Z_x \rangle \operatorname{Res}_x(f_x dg_x)$$

where $\langle\,,\rangle$ denotes the normalized Killing form of $\mathfrak{g}$. This is called an *affine Lie algebra*. Since the sum of the residues of a rational function on C is zero, the inclusion $\mathfrak{g}(U) \subset \mathfrak{g}_X$ composed with the obvious linear map $\mathfrak{g}_X \subset \hat{\mathfrak{g}}_X$ is a homomorphism of Lie algebras. Now fix a positive integer l (the *level*) and choose an irreducible highest weight representation H of $\hat{\mathfrak{g}}_X$ on which the generator of the center (defined by the above cocycle) acts as scalar multiplication by l. (Perhaps we should remark that we have a natural embedding of $\hat{\mathfrak{g}}_X$ in a product of affine Lie algebras $\prod_{x \in X} \hat{\mathfrak{g}}_x$ and that each of these factors appears as a subalgebra; so a representation H as above is tantamount to giving for each $x \in X$ an irreducible standard highest (integral) weight representation H_x of $\hat{\mathfrak{g}}_x$ of level l; H is then the tensor product of these.) The associated *conformal block* is by definition the dual of the space of $\mathfrak{g}(U)$-coinvariants of H: $V := (H_{\mathfrak{g}(U)})^*$. This is a finite dimensional space whose associated projective space $\mathbb{P}(V)$ only depends on the isomorphism class of H. If a decoration is given, then this is even true for V itself and so we have a vector bundle $\mathcal{V}$ over $\mathcal{M}_{g,n}$. A remarkable feature of this bundle is that it comes with a natural flat connection, in other words, it naturally underlies a local system.

Now suppose that we have a cover $\pi : (\tilde{C}, \tilde{X}, \tilde{v}) \to (C, X, v)$ of decorated curves and put $\tilde{U} := \pi^{-1}U$. In order to pull back the conformal block V on U to one on $\tilde{U}$, we must manufacture a standard representation $\tilde{H}$ of $\mathfrak{g}_{\tilde{X}}$ out of H. There is an obvious choice for this: induce H to a representation of $\mathfrak{g}_{\tilde{X}}$ (via the inclusion $\mathfrak{g}_X \subset \mathfrak{g}_{\tilde{X}}$); it is a highest weight representation which has a maximal standard quotient—this is our $\tilde{H}$. There is a natural map $H_{\mathfrak{g}(U)} \to \tilde{H}_{\mathfrak{g}(\tilde{U})}$. This gives rise to a vector bundle homomorphism $\tilde{\mathcal{V}} \to \mathcal{V}$ over $\mathcal{M}(f)$. This bundle map can be shown to be a homomorphism of local systems.

There is also a direct image construction, which is simply gotten by restricting $\tilde{H}$ to $\mathfrak{g}(U)$. This leads to a vector bundle homomorphism $\mathcal{V} \to \tilde{\mathcal{V}}$ over $\mathcal{M}(f)$ which is also a homomorphism of local systems.

5.2 The Hopf product and its algebro-geometric incarnation

If S_i, $i = 1,2$, are two oriented connected, compact surfaces with nonempty boundary, then an embedding of their disjoint union in an oriented connected compact surface S defines a homomorphism $\Gamma(S_1) \times \Gamma(S_2) \to \Gamma(S)$ and in the stable range this defines the Hopf product $\bullet$ on $H_\bullet^{\mathbb{Z}}(\Gamma_\infty)$. In particular, an embedding of n disjoint copies of S_1 in S, defines a homomorphism $\Gamma(S_1)^n \to \Gamma(S)$, which composed with the diagonal embedding $\Gamma(S_1) \to \Gamma(S_1)^n$, induces on $H_\bullet^{\mathbb{Z}}(\Gamma_\infty)$ the Adams operator ψ_n in the stable range.

Now let $f : \tilde{S} \to S$ be a connected covering of degree d. Denote by $\Gamma(f)$ the group of pairs $(\tilde{h}, h) \in \Gamma(\tilde{S}) \times \Gamma(S)$ with $f\tilde{h} = hf$. The projection $\Gamma(f) \to \Gamma(S)$ has finite kernel and maps onto a subgroup of finite index. So the rational homology of $\Gamma(S)$ appears as a direct summand of the rational homology of $\Gamma(f)$. The restriction of the second projection to this summand defines a correspondence $T[f] : H_\bullet(\Gamma(S)) \to H_\bullet(\Gamma(\tilde{S}))$ that takes counit to counit. Choose a connected component $\tilde{S}_i$ of $f^{-1}S_i$. Then the degree d_i of the covering $f_i : \tilde{S}_i \to S_i$ will divide d and $f^{-1}S_i$ consists of d/d_i copies of $\tilde{S}_i$. We have defined similarly $T[f_i] : H_\bullet(\Gamma(S_i)) \to H_\bullet(\Gamma(\tilde{S}_i))$. It is now clear that for $a_i \in H_{r_i}(\Gamma(S_i)) \cong H_{r_i}(\Gamma_\infty)$ in the stable range with respect to $g(S_i)$, we have the product formula

$$T[f](a_1 \bullet a_2) = \psi_{d/d_1} T[f_1](a_1) \bullet \psi_{d/d_2} T[f_2](a_2). \tag{5.2}$$

Now recall that $T(f)$ is the average of all $T[hf]$, where h runs over a system of representatives of cosets of the image of $\Gamma(f)$ in $\Gamma(S)$. Since h will in general not respect the subsurfaces S_i, we may not, in the above expression, replace brackets by parentheses.

The Hopf product is realized inside the Deligne-Mumford compactification in the following way. Out of two smooth once-pointed complex projective curves (C_1, x_1) and (C_2, x_2) of genus g_1 resp. g_2, we can fabricate a stable once-pointed curve (C, x) of genus $g = g_1 + g_2$ by attaching them both to $\mathbb{P}^1$: identify x_1 resp. x_2 with 0 resp. ∞ and taking for x the image of $1 \in \mathbb{P}^1$. This clearly defines a map $k : \mathcal{M}_{g'}^1 \times \mathcal{M}_{g_2}^1 \to \overline{\mathcal{M}}_{g,1}$, simply use the standard affine differential of $\mathbb{P}^1$ to decorate x. The image of k is of complex codimension two and lies in the locus where the boundary divisor has exactly two branches. It has a normal bundle ν_k in the orbifold sense, the normal space to the point defined by (C, x) being naturally identified with the direct sum $T_{x_1}C_1 \oplus T_{x_2}C_2$. Each summand defines normal vectors pointing along a branch of the boundary divisor and so a vector $(t_1, t_2) \in T_{x_1}C_1 \oplus T_{x_2}C_2$ points to the interior if and only if both t_1 and t_2 are nonzero. In other words, if we start with stable once decorated curves (C_i, v_i), then we not only get a boundary point of $\mathcal{M}_g^1$, but a normal vector pointing to the interior as well. Let $E(k)$ denote the normal bundle of the stratum of $\overline{\mathcal{M}}_{g,1}$ containing the image of k, and let $E'(k)$ be the $(\mathbb{C}^\times)^2$-subbundle of normal vectors pointing towards the interior. So we just defined a lift of k,

$$\tilde{k} : \mathcal{M}_{g_1,1} \times \mathcal{M}_{g_2,1} \to E'(k).$$

We should think of $E'(k)$ as the intersection with $\mathcal{M}_{g,1}$ of a regular neighborhood of the stratum containing the image of k. In particular, there is a natural map on homology $H_\bullet(E'(k)) \to H_\bullet(\mathcal{M}_{g,1})$. Its composite with $\tilde{k}$,

$$H_\bullet(\mathcal{M}_{g_1,1}) \otimes H_\bullet(\mathcal{M}_{g_2,1}) \to H_\bullet(\mathcal{M}_{g,1}),$$

realizes the Hopf product. A geometric picture of its behaviour with respect to correspondences is obtained by composing this map with a correspondence $\overline{T}(f) : \overline{\mathcal{M}}_{g,1} \rightrightarrows \mathcal{S}_m \backslash \overline{\mathcal{M}}_{h,m}$.

5.3   The stable Hecke operator attached to a finite group

Next consider the simplest case when S_1, S_2 and S have a single boundary circle and S minus the union of the interiors of S_1 and S_2 is a three holed sphere P. Choose a base point $* \in \partial S$ and choose a path in P connecting $*$ with a point $*_i \in \partial S_i$ so that $\pi_1(S_i, *_i)$ can be identified with a subgroup of $\pi(S, *)$. Notice that $\pi(S, *)$ is the free product of these two subgroups. We now fix a finite group G of order d. Every homomorphism $u : \pi_1(S, *) \to G$ defines a G-covering $f_u : \tilde{S}_u \to S$ whose degree d_u is the order of image(u). Put

$$\mathcal{T}_{S,G} := |\operatorname{Hom}(\pi_1(S, *), G)|^{-1} \sum_{u \in \operatorname{Hom}(\pi_1(S, *), G)} \psi_{d/d_u} T[f_u], \qquad (5.3)$$

viewed as homomorphism from the stable range homology of $\Gamma(S)$ to the stable range homology of $\Gamma(\tilde{S})$. So this is a weighted average of the operators $\psi_{d/d_u} T[f_u]$. In particular, $\mathcal{T}_{S,G}$ sends counit to counit. Our reason for introducing these maps is their nice behavior with respect to the Hopf product:

Proposition 5.3. *We have* $\mathcal{T}_{S,G}(1) = 1$ *and if* $a_i \in H_{s_i}(\Gamma(S_i))$ *is in the stable range, then* $\mathcal{T}_{S,G}(a_1 \bullet a_2) = \mathcal{T}_{S_1,G}(a_1) \bullet \mathcal{T}_{S_2,G}(a_2)$.

Proof. The first assertion is clear and included for the purpose of reference only. Since $\pi(S, *)$ is the free product of $\pi_1(S_1, *_1)$ and $\pi_1(S_2, *_2)$, we have a natural bijection between $\operatorname{Hom}(\pi_1(S, *), G)$ and $\operatorname{Hom}(\pi_1(S_1, *_1), G) \times \operatorname{Hom}(\pi_1(S_2, *_2), G)$. The second assertion now follows from the product formula (5.2). $\square$

Corollary 5.4. *The operator* $\mathcal{T}_{S,G}$ *in* $H_s(\Gamma_\infty)$ *for* $s \leq cg(S)$ *is independent of* S *and the resulting endomorphism* $\mathcal{T}_G$ *of* $H_\bullet(\Gamma_\infty)$ *is an algebra homomorphism.*

Proof. By taking $a_2 = 1$ in the above proposition, we see that $\mathcal{T}_{S,G}$ and $\mathcal{T}_{S_1,G}$ act identically in the stable range. The first statement follows and the second is clear. $\square$

I do not know whether $\mathcal{T}_G$ preserves the coproduct. By taking G abelian we get Proposition 4.1.

5.4   Stable covers and the stable Hecke operators they define

Remember that $S_{g,1}$ is a connected oriented compact surface of genus g with a single boundary component. Choose a base point in this boundary component and let $\pi_{g,1}$ denote the fundamental group of $S_{g,1}$ relative this base point. Let be given for every g a cofinite subgroup of $\pi_{g,1}$. We say that the collection $(I(g) \subset \pi_{g,1})_g$ is *stable* if for each g,

(i) $I(g)$ is a $\Gamma(S_{g,1})$-invariant subgroup of $\pi_{g,1}$ of finite index and

(ii) there exists a sense preserving embedding of $S_{g,1}$ in $S_{g+1,1}$ mapping base point to base point such that the preimage of $I(g+1)$ is equal to $I(g)$.

The homomorphisms $\pi_{g,1} \to \pi_{h,1}$, $h \geq g$, that arise from embeddings of $S_{g,1}$ in $S_{h,1}$ lie in a single $\Gamma(S_{h,1})$-orbit and so for a stable sequence $(I(g))_g$, the preimage of $I(h)$ under any such homomorphisms is equal to $I(g)$. (A natural subgroup of $\pi_{g,n}$ of finite index is now defined for every n: embed $S_{g,n}$ in some $S_{h,1}$ such that base point goes to base point and the closure of every connected component of the complement $S_{h,1} - S_{g,n}$ meets $S_{g,n}$ in a single boundary circle. Such embeddings belong to a single homotopy class and so the preimage of $I(g)$ in $\pi_{g,n}$ is independent of choices.) We may think of the sequence $(I(g))_g$ as being given by a subgroup I of the fundamental group of a connected pointed surface $(S, *)$ of infinite genus with the property that

(i) I is invariant under all compactly supported mapping classes of $(S, *)$ and

(ii) I has cofinite intersection with the fundamental group of any compact sub-surface $S' \subset S$ containing the base point.

Geometrically, a stable sequence amounts to giving a finite $\Gamma(S_{g,1})$-invariant covering $pr_g : \tilde{S}_{g,1} \to S_{g,1}$ for every g such that the pull-back of pr_h under an embedding $S_{g,1} \to S_{h,1}$ has each connected component $S_{g,1}$-isomorphic to $pr_g : \tilde{S}_{g,1} \to S_{g,1}$. This can also be described by a covering $pr : \tilde{S} \to S$ of our infinite genus surface S satisfying the two properties corresponding to the ones above. We call $(pr_g)_g$ a *stable sequence of coverings*. An intersection of two stable sequences is again stable.

Example 5.5. Let d be a positive integer and take for $I(g)$ the kernel of the homomorphism $\pi_{g,1} \to H_1(\pi_{g,1}; \mathbb{Z}/d)$. This defines a stable sequence.

Example 5.6. Fix a finite group G and define $I_G(g)$ as the intersection of the kernels of all group homomorphisms $\pi_{g,1} \to G$. This is certainly an invariant group. Since $\pi_{g,1}$ is finitely generated there are only finitely many homomorphisms from $\pi_{g,1}$ to the finite group G and so $I_G(g)$ is of finite index in $\pi_{g,1}$. The sequence $(I_G(g))_g$ is stable: since every homomorphism $\pi_{g,1} \to G$ can be extended to $\pi_{g+1,1}$, the pull-back of $I_G(g+1)$ is equal to $I_G(g)$. Notice that for $G = \mathbb{Z}/d$ we recover the previous example.

This example also shows that the stable sequences define a system of subgroups of $\pi_{g,1}$ of finite index that is cofinal among all such subgroups. For if π is any subgroup of $\pi_{g,1}$ of finite index, then the intersection of all its conjugates is a normal subgroup $N \subset \pi_{g,1}$ of finite index contained in π. If we put $G := \pi_{g,1}/N$, then clearly, $I_G(g) \subset N$.

Let $(I(g))_g$ and $(pr_g : \tilde{S}_{g,1} \to S_{g,1})_g$ be as above. Then $\tilde{S}_{g,1}$ is isomorphic to some $S_{g',n'}$ and so pr_g defines a correspondence that is almost a morphism $T_{I(g)} : \mathcal{M}_{g,1} \rightrightarrows \mathcal{M}_{g',n'}$: the ambiguity lies only in the decoration, so that its composite with $\mathcal{M}_{g',n'} \to \mathcal{M}_{g'}$ is a morphism indeed. In particular, it acts in the stable range as a cohomomorphism. Now let $S_{g_1,1}$ and $S_{g_2,1}$ be disjointly embedded in $S_{g,1}$ in a sense preserving way. If $d(g)$ denotes the index of $I(g)$ in $\pi_{g,1}$, then the restriction of pr_g to $S_{g_i,1}$ consists of $d(g)/d(g_i)$ copies of pr_{g_i}. So if $a_i \in H_{s_i}(\Gamma_\infty)$ with s_i in the stable range with respect to g_i, then we have a product formula just as in (5.2)

$$T_{I(g)}(a_1 \bullet a_2) = \psi_{d(g)/d(g_1)} T_{I(g_1)}(a_1) \bullet \psi_{d(g)/d(g_2)} T_{I(g_2)}(a_2). \tag{5.4}$$

From this we derive:

Proposition 5.7. *The action of* $\psi^{-1}_{[\pi_{g,1}:I(g)]} T_{I(g)}$ *on* $H_s(\Gamma_\infty)$ *is (for g in the stable range with respect to s) independent of g. The resulting endomorphism S_I of* $H_\bullet(\Gamma_\infty)$ *is a Hopf algebra endomorphism.*

Proof. Apply $\psi^{-1}_{d(g)}$ to formula (5.4). Substituting $a_2 = 1$ gives $\psi^{-1}_{d(g_1)} T_{I(g_1)} = \psi^{-1}_{d(g)} T_{I(g)}$. This proves the first assertion. The formula also shows that S_I is an algebra homomorphism. We already noticed that it is a cohomomorphism. $\square$

Proposition 4.4 follows if this proposition is applied to the stable sequence of Example 5.5.

Problem 5.8. I do not know whether the Hopf algebra endomorphism S_I is in fact an automorphism, let alone whether it is the identity (by 3.3 it is so on the tautological subalgebra).

6   Correspondences Acting on a Lie Algebra

We may think of $S_n \backslash \mathcal{M}_g^n$ as the moduli space of connected smooth affine curves of genus g with $n > 0$ punctures. Let us collect those with prescribed (negative) Euler characteristic $-e < 0$:

$$\mathcal{A}_e := \bigsqcup_{2g-2+n=e, n>0} S_n \backslash \mathcal{M}_g^n.$$

If we drop the condition that the curves be connected, we get

$$\mathcal{B}_e := \bigsqcup_{k_1 + 2k_2 + \cdots e k_e = e} \mathcal{A}_1^{k_1} \times \mathcal{A}_2^{k_2} \times \cdots \times \mathcal{A}_e^{k_e}.$$

According to Kontsevich, the cohomology of $\mathcal{A}_e$ has a remarkable interpretation. Consider $\mathbb{Q}^{2r}$ with the standard symplectic element $\omega_r := e_1 \wedge e_2 + \cdots + e_{2r-1} \wedge e_{2r} \in \wedge^2(\mathbb{Q}^{2r})$. Regard $\mathbb{Q}^{2r}$ as a graded vector space which is homogeneous of degree -1 and consider the derivations of the tensor algebra of $\mathbb{Q}^{2r}$ which kill ω_r and have degree ≤ 0. This is a graded Lie algebra which we denote by $\mathfrak{g}_r$. Its degree zero summand $\mathfrak{g}_{r,0}$ can be identified with the symplectic Lie $\mathbb{Q}$-algebra $\mathfrak{sp}_{2r}(\mathbb{Q})$. Then is defined the relative cohomology $H^\bullet(\mathfrak{g}_r, \mathfrak{g}_{r,0})$. Then the relative cohomology $H^\bullet(\mathfrak{g}_r, \mathfrak{g}_{r,0})$ is defined. The grading of $\mathfrak{g}_r$ defines one on each $H^s(\mathfrak{g}_r, \mathfrak{g}_{r,0})$ and the latter has degrees ≥ 0 only. The natural embeddings $(\mathfrak{g}_r, \mathfrak{g}_{r,0}) \subset (\mathfrak{g}_{r+1}, \mathfrak{g}_{r+1,0})$ of graded pairs of Lie algebras define graded maps $H^s(\mathfrak{g}_{r+1}, \mathfrak{g}_{r+1,0}) \to H^s(\mathfrak{g}_r, \mathfrak{g}_{r,0})$. These can be proved to stabilize and the limit $H^\bullet(\mathfrak{g}_\infty, \mathfrak{g}_{\infty,0})$ is naturally a bigraded Hopf algebra (the coproduct is induced by the obvious maps $(\mathfrak{g}_{r_1}, \mathfrak{g}_{r_1,0}) \times (\mathfrak{g}_{r_2}, \mathfrak{g}_{r_2,0}) \to (\mathfrak{g}_{r_1+r_2}, \mathfrak{g}_{r_1+r_2,0})$). Let $H^\bullet_{\mathrm{pr}}(\mathfrak{g}_\infty, \mathfrak{g}_{\infty,0})$ denote its primitive part. Kontsevich [7] (see also [3]) proves that we have a canonical isomorphism

$$H_k(\mathcal{A}_e) \cong H^{2e-k}_{\mathrm{pr}}(\mathfrak{g}_\infty, \mathfrak{g}_{\infty,0})_{2e}.$$

This is equivalent to saying that we have canonical isomorphisms

$$H_k(\mathcal{B}_e) \cong H^{2e-k}(\mathfrak{g}_\infty, \mathfrak{g}_{\infty,0})_{2e}$$

whose direct sum is an isomorphism of Hopf algebras. (The Hopf algebra structure on the homology of the disjoint union of the $\mathcal{B}_e$'s comes from 'taking disjoint union of curves'.)

Via this isomorphism correspondences yield operations in $H^\bullet_{\mathrm{pr}}(\mathfrak{g}_\infty, \mathfrak{g}_{\infty,0})_\bullet$. For instance, if we are given for every connected oriented surface of Euler characteristic $-e$ a connected covering of degree d up to homeomorphism, then this determines a correspondence $H_\bullet(\mathcal{A}_e) \to H_\bullet(\mathcal{A}_{de})$, and so a linear map

$$t : H^\bullet_{\mathrm{pr}}(\mathfrak{g}_\infty, \mathfrak{g}_{\infty,0})_{2e} \to H^{\bullet+2(d-1)e}_{\mathrm{pr}}(\mathfrak{g}_\infty, \mathfrak{g}_{\infty,0})_{2de}.$$

This map increases both grade and dimension by $2(d-1)e$ and so has the formal appearance of a dth power operation.

Problem 6.1. Give an interpretation of the maps t (for some natural choices of covers, say) in terms of the Lie pair $(\mathfrak{g}_\infty, \mathfrak{g}_{\infty,0})$.

We believe that such an interpretation could be very useful. Perhaps this is also the place to remark that to the best of the our knowledge none of the finer structure that exists on the homology of $\mathcal{A}_e$ (the coproduct, the Hodge decomposition, ...) has been transcribed to the Lie algebra side. And now that we are at it, a similar interpretation for the homology of the Deligne-Mumford compactification $\overline{\mathcal{A}}_e$ is conspicuously missing.

Bibliography

[1] A. Beauville: *Conformal blocks, fusion rules and the Verlinde formula*, in Proc. Hirzebruch 65 Conference on Algebraic Geometry, Israel Math. Conf. Proc. 9 (1996), 75–96.

[2] I. Biswas, S. Nag, D. Sullivan: *Determinant bundles, Quillen metrics and Mumford isomorphisms over the universal commensurability Teichmüller space*, Acta Math. 176 (1996), 145–169.

[3] R. Hain and E. Looijenga: *Mapping Class Groups and Moduli Spaces of Curves*, in Algebraic Geometry—Santa Cruz 1995, Proc. Symp. Pure Math. 62, Part 2 (1997), 97–142.

[4] J. Harer: *Stability of the homology of the mapping class groups of orientable surfaces*, Ann. of Math. 121 (1985), 215–249.

[5] J. Harer: *Improved stability for the homology of the mapping class groups of surfaces*, Duke University preprint, 1993 (http://www.math.duke.edu/preprints/).

[6] J. Harris and D. Mumford: *On the Kodaira dimension of the moduli space of curves*, Invent. Math. 67 (1982), 23–86.

[7] M. Kontsevich: *Feynman diagrams and low-dimensional topology*, in: Proc. First Eur. Math. Congr. at Paris (1992), Vol. II, 97–121, Birkhäuser Verlag, Basel (1994).

[8] K. Ueno: *Introduction to Conformal Field Theory with Gauge Symmetries*, in: Geometry and Physics, Proc. of a Conf. at Aarhus Univ. (1995), 603–745. Lecture Notes in Pure and Appl. Math. Vol. 184, Marcel Dekker, Inc., New York (1997).

Fields, Strings, Matrices and Symmetric Products

Robbert Dijkgraaf

Abstract

In these notes we review the role played by the quantum mechanics and sigma models of symmetric products in the quantization of quantum field theories, string theory and matrix theory.

1 Introduction

For more than a decade now, string theory has been a significant, continuous influence in mathematics, ranging through fields as diverse as algebraic geometry and representation theory. However, it is fair to say that most of these applications concerned the so-called first-quantized formulation of the theory, the formulation that is used to describe the propagation of a single string. In contrast with point-particle theories, in string theory the first-quantized theory is so powerful because it naturally can be extended to also describe the perturbative interactions of splitting and joining of strings by means of Riemann surfaces of general topology. Study of these perturbative strings has led to series of remarkable mathematical developments, such as representation theory of infinite-dimensional Lie algebras, mirror symmetry, quantum cohomology and Gromov-Witten theory.

The second-quantized formalism, what is sometimes also referred to as string field theory, has left a much smaller mathematical imprint. Although there is a beautiful geometrical and algebraic structure of perturbative closed string field theory, developed mainly by Zwiebach [1], which is built on deep features of the moduli space of Riemann surfaces, it is very difficult to analyze, perhaps because it is intrinsically perturbative. Yet, in recent years it has become clear that the non-perturbative mathematical structure of string theory is even richer than the perturbative one, with even bigger symmetry groups—the mysterious U-duality groups [2]. The appearance of D-branes [3] and an eleven-dimensional origin in the form of M-theory [4] can only be properly understood from a second-quantized point of view.

At present there is only one candidate for a fundamental description of non-perturbative string theory, which is matrix theory [5]. In matrix theory an important role is played by non-abelian gauge fields, and the strings and conformal field theory only emerge in a certain weak coupling limit. We will not review much of matrix theory in these notes but refer to for example [6, 7, 8, 9]. Important is that matrix theory makes direct contact with the second-quantized theory, indeed Fock spaces are one of the ubiquitous ingredients, and much of the notes will focus on this correspondence, also reviewing the work of [10, 11].

1.1 Symmetric products and Gromov-Witten invariants

Let us briefly explain what the possible relations are of the matters discussed in this paper with Gromov-Witten theory and the moduli space of curves.

String theory can be considered as a two-parameter deformation of classical geometry. The first deformation is related to the fact that we study a manifold X through its free loop space $\mathcal{L}X$. In physical terms this corresponds to going from point particles to strings. Quantization of a (supersymmetric) point particle in quantum mechanics associates to a manifold X a Hilbert space $\mathcal{H} = \Omega^*(X)$ together with a self-adjoint operator, the Hamiltonian, given by the Laplacian. Much of algebraic topology and Hodge theory can be recovered in this way — a point forcefully demonstrated by the work of Witten.

In string theory this Hilbert space is replaced by the space of semi-infinite differential forms on the loop space. This amounts to studying the Floer homology of X. This quantization depends on the volume of X through a parameter α' as explained in more detail in section 3. Roughly, from a path-integral point of view, string theory concerns the sum over all maps of a Riemann surface Σ into X weighted by the action

$$A_\Sigma(\alpha') = \sum_{\Sigma \to X} e^{-Area(\Sigma)/\alpha'} \qquad (1.1)$$

In the limit $\alpha' \to 0$ this sum is localized on the constant maps, and classical geometry is recovered. From the point of view of Gromov-Witten theory or quantum cohomology, the deformation of the cup product on $H^*(X)$ is in terms of the formal variable $q = e^{-1/\alpha'}$. Since such terms do not contribute in a power series expansion in α', they are called world-sheet instantons.

The second deformation parameter concerns the genus expansion. Gromov-Witten theory produces invariants of X defined using Riemann surfaces of a fixed topology, say genus g. Let us denote such invariants generically as $A_g(\alpha')$. In string theory one introduces the string coupling constant λ by considering formal sums of the form

$$\sum_{g \geq 0} \lambda^{2g-2} A_g(\alpha'). \tag{1.2}$$

Essentially because the volume of the moduli space of Riemann surfaces $\mathcal{M}_g$ grows as $(2g)!$ these expressions are generically not convergent, and can only serve as *asymptotic* expansions of a so-called non-perturbative object $A(\lambda, \alpha')$. Finding the objects $A(\lambda, \alpha')$ and their fundamental properties is the final aim of string theory. As we have seen in Kontsevich's proof of the Witten conjectures, and indeed in the formulation of these conjectures using integrable hierarchies, there are natural functions $A(\lambda)$ that have the property that their expansion can be done in a sum over topologies of surfaces.

The parameter α' that controls the size of the string can be considered as Planck's constant for the two-dimensional conformal field theory (CFT) on the world-sheet Σ of the string. The parameter λ must be seen as Planck's constant for the quantum theory induced on the space-time manifold X. Understanding the λ dependence makes it necessary to study not one but many strings, since the non-trivial topology of the Riemann surfaces corresponds to space-time diagrams where strings join and split. In physical terms, we have to consider the second-quantization. This leads one to study not the Gromov-Witten theory on the manifold X but on the symmetric product X^N/S_N, which is the aim of this paper. It therefore sets the stage of what ultimately could be a non-perturbative Gromov-Witten theory.

1.2 Hamilton vs Lagrange: representation theory and automorphic forms

One of the most remarkable insights provided by string theory, or more properly conformal field theory, is the natural explanation it offers of the modular properties of the characters of affine Kac-Moody algebras, Heisenberg algebras, and other infinite-dimensional Lie algebras. At the heart of this explanation—and in fact of much of the applications of field theory in mathematics—lies the equivalence between the Hamiltonian and Lagrangian formulation of quantum mechanics and quantum field theory.

This equivalence roughly proceeds as follows (see also [15]). In the Hamiltonian formulation one considers the quantization of a two-dimensional conformal field theory on a space-time cylinder $\mathbb{R} \times S^1$. The basic object is the loop space $\mathcal{L}X$ of maps $S^1 \to X$ for some appropriate target space X. The infinite-dimensional Hilbert space $\mathcal{H}$ that forms the representation of the algebra of quantum observables is then typically obtained by quantizing the loop space $\mathcal{L}X$. This Hilbert space carries an obvious S^1 action, generated by the momentum operator P that rotates the loop, and the character of the representation is defined as

$$\chi(q) = \mathrm{Tr}_{\mathcal{H}} q^P \tag{1.3}$$

with $q = e^{2\pi i \tau}$ and τ in the complex upper half-plane $\mathbb{H}$. The claim is that these characters are always some kind of modular forms. From the representation theoretic point of view it is not at all clear why there should be a natural action of the modular group $SL(2,\mathbb{Z})$ acting on τ by linear fractional transformations. In particular the transformation $\tau \to -1/\tau$ is rather mysterious.

In the Lagrangian formulation, however, the character $\chi(q)$, or more properly the partition function, is computed by considering the quantum field theory on a Riemann surface with topology of a two-torus $T^2 = S^1 \times S^1$, *i.e.*, an elliptic curve with modulus τ. The starting point is the path-integral over all maps $T^2 \to X$. Since we work with an elliptic curve, the modularity is built in from the start. The transformation $\tau \to -1/\tau$ simply interchanges the two S^1's. Changing from the Hamiltonian to the Lagrangian perspective, we understand the appearance of the modular group $SL(2,\mathbb{Z})$ as the 'classical' automorphism group of the two-torus. This torus is obtained by gluing the two ends of the cylinder $S^1 \times \mathbb{R}$, which is the geometric equivalent of taking the trace. Note that in string theory this two-torus typically plays the role of a *world-sheet*.

In second-quantized string theory we expect a huge generalization of this familiar two-dimensional story. The operator algebras will be much bigger (typically, generalized Kac-Moody algebras) and also the automorphism groups will not be of a classical form, but will reflect the 'stringy' geometry at work. An example we will discuss in great detail in these notes is the quantization of strings on a space-time manifold of the form

$$M = \mathbb{R} \times S^1 \times X, \tag{1.4}$$

with X a compact simply-connected Riemannian manifold. Quantization leads again to a Hilbert space $\mathcal{H}$, but this space carries now at least two circle actions.

First, we have again a momentum operator P that generates the translations along the S^1 factor. Second, there is also a winding number operator W that counts how many times a string is wound around this circle. It labels the connected components of the loop space $\mathcal{L}M$. A state in $\mathcal{H}$ with eigenvalue $W = m \in \mathbb{Z}$ represents a string that is wound m times around the S^1. In this way we can define a two-parameter character

$$\chi(q,p) = \mathrm{Tr}_{\mathcal{H}} p^W q^P, \tag{1.5}$$

with $p = e^{2\pi i \sigma}$, $q = e^{2\pi i \tau}$, with both $\sigma, \tau \in \mathbb{H}$. We will see in concrete examples that these kind of expressions will be typically the character of a generalized Kac-Moody algebra and transform as automorphic forms.

The automorphic properties of such characters become evident by changing again to a Lagrangian point of view and computing the partition function on the compact manifold $T^2 \times X$. Concentrating on the T^2 factor, which now has an interpretation as a *space-time*, the string partition function carries a manifest T-duality symmetry group

$$SO(2,2,\mathbb{Z}) \cong SL(2,\mathbb{Z}) \times SL(2,\mathbb{Z}), \tag{1.6}$$

which is the 'stringy' automorphism group of T^2.

Let us explain briefly how this group acts on the moduli σ, τ. Since the string theory is not a conformal field theory, the partition function will depend both on the modulus τ of T^2 and on its volume g. Furthermore there is an extra dependence on a constant 2-form field $\theta \in H^2(T^2, \mathbb{R}/\mathbb{Z})$. These two extra data are combined in a second complex 'modulus' $\sigma = \theta + ig$. The T-duality group $SO(2,2,\mathbb{Z})$ will now act on the pair (σ, τ) by separate fractional linear transformations and the generalized character (1.5) will be some automorphic form for this group. Of course only the second $SL(2,\mathbb{Z})$ factor has a clear geometric interpretation. The first factor, that exchanges large and small volume $\sigma \to -1/\sigma$, has a complete stringy origin.

The appearance of the T-duality group $SO(2,2,\mathbb{Z})$ as a symmetry group of the two-torus is most simply explained by considering a single string. We are then dealing with the loop space $\mathcal{L}T^2$. If the torus is given by the quotient $\mathbb{R}^2/\Lambda$, with Λ a two-dimensional lattice, the momenta of such a string take value in the dual lattice

$$P = \oint \dot{x} \in \Lambda^*. \tag{1.7}$$

The winding numbers, that label the components of $\mathcal{L}T^2$, lie in the original lattice

$$W = \oint dx \in \Lambda. \tag{1.8}$$

Therefore the total vector $v = (W, P)$ can be seen as an element of the rank 4, signature (2,2), even, self-dual Narain lattice

$$v = (W, P) \in \Gamma^{2,2} = \Lambda \oplus \Lambda^*, \qquad v^2 = 2W \cdot P \tag{1.9}$$

The T-duality group appears now as the automorphism group of the lattice $\Gamma^{2,2}$. In the particular example we will discuss in detail, where the manifold X is a Calabi-Yau space, there will be an extra quantum number and the lattice will be enlarged to a signature (2,3) lattice. Correspondingly, the automorphism group will be given by $SO(3,2,\mathbb{Z}) \cong Sp(4,\mathbb{Z})$.

2   Particles, Symmetric Products and Fields

It is a well-known wise-crack that first-quantization is a mystery but second-quantization a functor. Indeed, for a free theory second quantization involves nothing more than taking symmetric products. We obtain the second-quantized Hilbert space from the first-quantized Hilbert space $\mathcal{H}$ as the free symmetric algebra $S\mathcal{H}$. Yet, recent developments in string theory (and in certain field theories that are naturally obtained as limits of string theories) have provided us with a fresh outlook on this familiar subject. In particular this new approach allow us to include interactions in new ways.

2.1   Second-quantization of superparticles

Let us start by considering a well-known case: a point-particle moving on a compact oriented Riemannian manifold X. The first-quantization 'functor' Q^1 of quantum mechanics assigns to each manifold X a Hilbert space $\mathcal{H}$ and a Hamiltonian H,

$$Q^1 : X \mapsto (\mathcal{H}, H). \tag{2.1}$$

As is well-known, in (bosonic) quantum mechanics the Hilbert space is given by the square-integrable functions on X, $\mathcal{H} = L^2(X)$, together with the positive-definite Hamiltonian $H = -\frac{1}{2}\Delta$, with Δ the Laplacian on X.

Supersymmetry adds anticommuting variables, and for the supersymmetric particle the Hilbert space is now the L^2-completion of the space of differential forms on X,

$$\mathcal{H} = \Omega^*(X) \tag{2.2}$$

On this space we can realize the elementary $N = 2$ supersymmetry algebra

$$[Q, Q^*] = -2H \tag{2.3}$$

by the use of the supercharge or differential $Q = d$ and its adjoint Q^*. The spectrum of the Hamiltonian is encoded in the partition function

$$Z(X; q, y) = \mathrm{Tr}_{\mathcal{H}}(-1)^F q^H y^F \tag{2.4}$$

with fermion number F given by the degree of the differential form.

Of particular interest is the subspace $\mathcal{V} \subset \mathcal{H}$ of supersymmetric ground states, that satisfy $Q\psi = Q^*\psi = 0$ and therefore also $H\psi = 0$. These zero-energy wave functions are represented by harmonic differential forms

$$\mathcal{V} = \mathrm{Harm}^*(X) \cong H^*(X). \tag{2.5}$$

We can compute the weighted number of ground states by the Witten index, which defines a regularized superdimension[1] of the Hilbert space

$$\mathrm{sdim}\,\mathcal{H} = \mathrm{Tr}_{\mathcal{H}}(-1)^F = Z(X; q, 1) \tag{2.6}$$

Since this expression does not depend on q, the Witten index simply equals the Euler number of the space X

$$\mathrm{Tr}_{\mathcal{H}}(-1)^F = \mathrm{sdim}\,\mathcal{V} = \sum_k (-1)^k \dim H^k(X) = \chi(X). \tag{2.7}$$

Note that here we consider $H^*(X)$ as a graded vector space generated by b^+ even generators and b^- odd generators with $\chi(X) = b^+ - b^-$.

---

[1] For a graded vector space $V = V^+ \oplus V^-$ with even part V^+ and odd part V^-, we define the superdimension as $\mathrm{sdim}\,V = \dim V^+ - \dim V^-$, and, more generally, the supertrace of an operator a acting on V as $\mathrm{sTr}_V(a) = \mathrm{Tr}_{V^+}(a) - \mathrm{Tr}_{V^-}(a)$. So we have $\mathrm{sdim}\,V = \mathrm{sTr}_V 1 = \mathrm{Tr}_V(-1)^F$. Here the Witten index operator $(-1)^F$ is defined as $+1$ on V^+ and -1 on V^-.

2.2   Second-quantization and symmetric products

The usual step of second-quantization now consists of considering a system of N of these (super)particles. It is implemented by taking the N-th symmetric product of the single particle Hilbert space $\mathcal{H}$

$$S^N \mathcal{H} = \mathcal{H}^{\otimes n}/S_N, \tag{2.8}$$

or more properly the direct sum over all N

$$\mathcal{Q}_2 : \mathcal{H} \to S\mathcal{H} = \bigoplus_{N \geq 0} S^N \mathcal{H}. \tag{2.9}$$

We now propose to reverse roles. Instead of taking the symmetric product of the Hilbert space of functions or differential forms on the manifold X (*i.e.*, the symmetrization of the quantized manifold), we will take the Hilbert space of functions or differential forms on the symmetric product $S^N X$ (*i.e.*, the quantization of the symmetrized manifold)

$$\mathcal{Q}^2 : X \to SX = \coprod_{N \geq 0} S^N X. \tag{2.10}$$

The precise physical interpretation of this role-reversing is the topic of these notes. It will appear later as a natural framework for the light-cone quantization of string theory and of a certain class of quantum field theories that are obtained as low-energy limits of string theories. We will be particularly interested to learn whether these operations commute (they will not)

$$\begin{array}{ccc}
X & \xrightarrow{\mathcal{Q}^1} & \mathcal{H} \\
\downarrow \mathcal{Q}^2 & & \downarrow \mathcal{Q}^2 \\
SX & \xrightarrow{\mathcal{Q}^1} & S\mathcal{H}
\end{array} \tag{2.11}$$

But first we have to address the issue that the symmetric space $S^N X$ is not a smooth manifold but an orbifold, namely the quotient by the symmetric group S_N on N elements,

$$S^N X = X^N/S_N. \tag{2.12}$$

We will first be interested in computing the ground states for this symmetric product, which we have seen are in general counted by the Euler number. Actually, the relevant concept will turn out to be the orbifold Euler number. Using this concept there is a beautiful formula that was first discovered by Göttsche [16] (see also [17, 18]) in the context of Hilbert schemes of algebraic surfaces, but which is much more generally valid in the context of orbifolds, as was pointed out by Hirzebruch and Höfer [19].

First some notation. It is well-known that many formulas for symmetric products take a much more manageable form if we introduce generating functions. For a general graded vector space we will use the notation

$$S_p V = \bigoplus_{N \geq 0} p^N S^N V \tag{2.13}$$

for the weighted formal sum of symmetric products. Note that for graded vector spaces the symmetrization under the action of the symmetric group S_N is always to be understood in the graded sense, *i.e.*, antisymmetrization for the odd-graded pieces. We recall that for an even vector space

$$\dim S_p V = \sum_{N \geq 0} p^N \dim S^N V = (1 - p)^{-\dim V} \tag{2.14}$$

whereas for an odd vector space

$$\mathrm{sdim}\, S_p V = \sum_{N \geq 0} (-1)^N p^N \dim {\bigwedge}^N V = (1 - p)^{\dim V} \tag{2.15}$$

These two formulas can be combined into the single formula valid for an arbitrary graded vector space that we will use often[2]

$$\mathrm{sdim}\, S_p V = (1 - p)^{-\mathrm{sdim}\, V} \tag{2.16}$$

Similar we introduce for a general space X the 'vertex operator'

$$S_p X = \text{'exp}\, pX\text{'} = \coprod_{N \geq 0} p^N S^N X. \tag{2.17}$$

Using this formal expression the formula we are interested in reads (see also [20])

   Theorem 1 [16, 19] — *The orbifold Euler numbers of the symmetric products $S^N X$ are given by*

$$\chi_{orb}(S_p X) = \prod_{n > 0} (1 - p^n)^{-\chi(X)}. \tag{2.18}$$

2.3   The orbifold Euler character

The crucial ingredient in Theorem 1 is the orbifold Euler character, a concept that is very nicely explained in [19]. Here we give a brief summary of its definition.

   Suppose a finite group G acts on a manifold M. In general this action will not be free and the space M/G is not a smooth manifold but an orbifold instead. The *topological* Euler number of this singular space, defined as for any topological space, can be computed as the alternating sum of the dimensions of the invariant piece of the cohomology,

---

[2] This can of course be generalized to traces of operators as $\mathrm{sTr}\,(S_p A) = \mathrm{sdet}(1 - pA)^{-1}$.

$$\chi_{top}(M/G) = \text{sdim}\, H^*(M)^G. \tag{2.19}$$

In the de Rham cohomology one can also simply take the complex of differential forms that are invariant under the G-action and compute the cohomology of the standard differential d. This expression can be computed by averaging over the group

$$\chi_{top}(M/G) = \frac{1}{|G|} \sum_{g \in G} \text{sTr}_{H^*(M)} g$$

$$= \frac{1}{|G|} \sum_{g \in G} g \, \boxed{}_{1} \tag{2.20}$$

Alternatively, using the Lefschetz fixed point formula, we can rewrite this Euler number as a sum of fixed point contributions. Let M^g denote the fixed point set of the element $g \in G$. (Note that for the identity $M^1 = M$.) Then we have

$$\chi_{top}(M/G) = \frac{1}{|G|} \sum_{g \in G} \text{sdim}\, H^*(M^g)$$

$$= \frac{1}{|G|} \sum_{g \in G} 1 \, \boxed{}_{g} \tag{2.21}$$

In the above two formulas we used the familiar string theory notation

$$h \, \boxed{}_{g} = \text{sTr}_{H^*(M^g)} h \tag{2.22}$$

for the trace of the group element h in the 'twisted sector' labeled by g. Note that the two expressions (2.20) and (2.21) for the topological Euler number are related by a 'modular S-transformation,' that acts as

$$g \, \boxed{}_{1} \rightarrow 1 \, \boxed{}_{g} \tag{2.23}$$

The *orbifold* Euler number is the proper equivariant notion. We see in a moment how it naturally appears in string theory. In the orbifold definition we remember that on each fixed point set M^g there is still an action of the centralizer or stabilizer subgroup C_g that consists of all elements $h \in G$ that commute with g. The orbifold cohomology is defined by including the fixed point loci M^g, but now taking only the contributions of the C_g invariants. That is, we have a sum over the conjugacy classes $[g]$ of G of the topological Euler character of these strata

$$\chi_{orb}(M/G) = \sum_{[g]} \chi_{top}(M^g/C_g). \tag{2.24}$$

Note that this definition always gives an integer, in contrast with other natural definitions of the Euler number of orbifolds. From this point of view the topological Euler number only takes into account the trivial class $g = 1$ (the 'untwisted sector'). If we use the elementary fact that $|[g]| = |G|/|C_g|$, we obtain in this way

$$\chi_{orb}(M/G) = \frac{1}{|G|} \sum_{g \in G} \text{sdim} \, H^*(M^g)^{C_g}$$

$$= \frac{1}{|G|} \sum_{g,h \in G, \, gh=hg} \text{sTr}_{H^\bullet(M^g)} h$$

$$= \frac{1}{|G|} \sum_{g,h \in G, \, gh=hg} h \, \boxed{}_{g} \qquad (2.25)$$

This definition is manifestly invariant under the 'S-duality' that exchanges h and g. We see that, compared with the topological definition, the orbifold Euler number contains extra contribution of the 'twisted sectors' corresponding to the non-trivial fixed point loci M^g. Using again Lefschetz's formula, it can be written alternatively in terms of the cohomology of the subspaces $M^{g,h}$ that are left fixed by both g and h as

$$\chi_{orb}(M/G) = \frac{1}{|G|} \sum_{g,h \in G, \, gh=hg} \text{sdim} \, H^*(M^{g,h}). \qquad (2.26)$$

It has been pointed out by Segal [21] that much of this and in particular the applications to symmetric products that we are about to give, find a natural place in equivariant K-theory. Indeed the equivariant K-group (tensored with $\mathbb{C}$) of a space M with a G action is isomorphic to

$$K_G(M) = \bigoplus_{[g]} K(M^g)^{C_g}. \qquad (2.27)$$

2.4   The orbifold Euler number of a symmetric product

We now apply the above formalism to the case of the quotient X^N/S_N. For the topological Euler number the result is elementary. We simply replace $H^*(X)$ by its symmetric product $S^N H^*(X)$. Since we take the sum over all symmetric products, graded by N, this is just the free symmetric algebra on the generators of $H^*(X)$, so that [22]

$$\chi_{top}(S_p X) = \text{sdim} \, S_p H^*(X) = (1-p)^{-\chi(X)} \qquad (2.28)$$

In order to prove the orbifold formula (2.18) we need to include the contributions of the fixed point sets. Thereto we recall some elementary facts about the symmetric group. First, the conjugacy classes $[g]$ of S^N are labeled by partitions $\{N_n\}$ of N, since any group element can be written as a product of elementary cycles (n) of length n,

$$[g] = (1)^{N_1}(2)^{N_2}\ldots(k)^{N_k}, \qquad \sum_{n>0} nN_n = N. \tag{2.29}$$

The fixed point set of such an element g is easy to describe. The symmetric group acts on N-tuples $(x_1,\ldots,x_N) \in X^N$. A cycle of length n only leaves a point in X^N invariant if the n points on which it acts coincide. So the fixed point locus of a general g in the above conjugacy class is isomorphic to

$$(X^N)^g \cong \prod_{n>0} X^{N_n}. \tag{2.30}$$

The centralizer of such an element is a semidirect product of factors S_{N_n} and $\mathbb{Z}_n$,

$$C_g = S_{N_1} \times (S_{N_2} \ltimes \mathbb{Z}_2^{N_2}) \times \ldots (S_{N_k} \ltimes \mathbb{Z}_k^{N_k}). \tag{2.31}$$

Here the factors S_{N_n} permute the N_n cycles (n), while the factors $\mathbb{Z}_n$ act within one particular cycle (n). The action of the centralizer C_g on the fixed point set $(X^N)^g$ is obvious: only the subfactors S_{N_n} act non-trivially giving

$$(X^N)^g/C_g \cong \prod_{n>0} S^{N_n} X. \tag{2.32}$$

We now only have to assemble the various components to compute the orbifold Euler number of $S^N X$:

$$
\begin{aligned}
\chi_{orb}(S_p X) &= \sum_{N \geq 0} p^N \chi_{orb}(S^N X) \\[2mm]
&= \sum_{N \geq 0} p^N \sum_{\substack{\{N_n\} \\ \sum nN_n = N}} \prod_{n>0} \chi_{top}(S^{N_n} X) \\[2mm]
&= \prod_{n>0} \sum_{N \geq 0} p^{nN} \chi_{top}(S^N X) \\[2mm]
&= \prod_{n>0} (1 - p^n)^{-\chi(X)} \tag{2.33}
\end{aligned}
$$

which concludes the proof of (2.18).

2.5   Orbifold quantum mechanics on symmetric products

The above manipulation can be extended beyond the computation of the Euler number to the actual cohomology groups. We will only be able to fully justify these definitions (because that's what it is at this point) from the string theory considerations that we present in the next section. For the moment let us just state that in particular cases where the symmetric product allows for a natural smooth resolution (as for the algebraic surfaces studied in [16] where the Hilbert scheme provides such a resolution), we expect the orbifold definition to be compatible with the usual definition in terms of the smooth resolution.

We easily define a second-quantized, infinite-dimensional graded Fock space whose graded superdimensions equal the Euler numbers that we just computed. Starting with the single particle ground state Hilbert space

$$\mathcal{V} = H^*(X) \tag{2.34}$$

we define it as the symmetric algebra of an infinite number of copies $\mathcal{V}^{(n)}$ graded by $n = 1, 2, \ldots$

$$\mathcal{F}_p = \bigotimes_{n>0} S_{p^n} \mathcal{V}^{(n)} = S\left(\bigoplus_{n>0} p^n \mathcal{V}^{(n)}\right). \tag{2.35}$$

Here $\mathcal{V}^{(n)}$ is a copy of $\mathcal{V}$ where the 'number operator' N is defined to have eigenvalue n, so that

$$\chi_{orb}(S_p X) = \text{Tr}_{\mathcal{F}} (-1)^F p^N = \prod_{n>0} (1 - p^n)^{-\chi(X)}. \tag{2.36}$$

We will see later that the degrees in $\mathcal{V}^{(n)}$ are naturally shifted by $(n-1)\frac{d}{2}$ with d the dimension of X, so that

$$\mathcal{V}^{(n)} \cong H^{*-(n-1)\frac{d}{2}}(X), \qquad n > 0. \tag{2.37}$$

Of course, this definition makes only good sense for even d, which will be the case since we will always consider Kähler manifolds.

This result can be interpreted as follows. We have seen that the fixed point loci consist of copies of X. These copies $X^{(n)}$ appear as the small diagonal inside $S^n X$ where all n points come together. If we think in terms of middle dimensional cohomology, which is particularly relevant for Kähler and hyperkähler manifolds, this result tells us that the middle dimensional cohomology of X contributes through $X^{(n)}$ to the middle dimensional cohomology of $S^n X$.

So, if we define the Poincaré polynomial as

$$P(X; y) = Z(X; 0, y) = \text{Tr}_{\mathcal{V}} (-1)^F y^F = \sum_{0 \le k \le d} (-1)^k y^k b_k(X), \tag{2.38}$$

then we claim that the orbifold Poincaré polynomials of the symmetric products $S^N X$ are given by

$$P_{orb}(S_p X; y) = \prod_{\substack{n>0 \\ 0 \le k \le d}} \left(1 - y^{k+(n-1)\frac{d}{2}} p^n\right)^{-(-1)^k b_k} \tag{2.39}$$

This is actually proved for the Hilbert scheme of an algebraic surface in [17, 18].

Although we will only be in a position to understand this well in the next section, we can also determine the full partition function that encodes the quantum mechanics on $S^N X$. Again the Hilbert space is a Fock space built on an infinite number of copies of the single particle Hilbert space $\mathcal{H}(X)$

$$\mathcal{H}_{orb}(S_p X) = \bigotimes_{n>0} S_{p^n} \mathcal{H}^{(n)}(X). \tag{2.40}$$

The contribution to the total Hamiltonian of the states in the sector $\mathcal{H}^{(n)}$ turns out to be scaled by a factor of n relative to the first-quantized particle, whereas the fermion number is shifted as before, so that

$$\mathcal{H}^{(n)} \cong \Omega^{*-(n-1)\frac{d}{2}}(X), \qquad H^{(n)} = -\tfrac{1}{2}\Delta/n. \tag{2.41}$$

To be completely explicit, let $\{h(m,k)\}_{m\geq 0}$ be the spectrum of H on the subspace $\Omega^k(X)$ of k-forms with degeneracies[3] $c(m,k)$, so that the single particle partition function reads

$$Z(X;q,y) = \mathrm{Tr}_{\mathcal{H}}(-1)^F y^F q^H = \sum_{\substack{n>0 \\ 0 \leq k \leq d}} c(m,k) y^k q^{h(m,k)}. \tag{2.42}$$

Then we have for the symmetric product (in the orbifold sense)

$$\begin{aligned}
Z_{orb}(S_p X; q,y) &= \mathrm{Tr}_{\mathcal{H}_{orb}(SX)}(-1)^F p^N q^H y^F \\
&= \prod_{\substack{n>0,\, m\geq 0 \\ 0 \leq k \leq d}} \left(1 - p^n q^{h(m,k)/n} y^{k+(n-1)\frac{d}{2}}\right)^{-c(m,k)}
\end{aligned} \tag{2.43}$$

In later sections we give a quantum field theory (QFT) interpretation of this formula.

3   Second-quantized Strings

The previous section should be considered just as a warming-up for the much more interesting case of string theory. We will now follow all of the previous steps again, going from a single quantized string to a gas of second-quantized strings. In many respects this construction — in particular the up to now rather mysterious orbifold prescription — is more 'canonical,' and all of the previous results can be obtained as a natural limiting case of the string computations.

---

[3] These degeneracies are consistently defined as superdimensions of the eigenspaces, so that $c(m,k) \leq 0$ for k odd, and $c(0,k) = (-1)^k b_k$.

3.1   The two-dimensional supersymmetric sigma model

In the Lagrangian formulation the supersymmetric sigma model that describes the propagation of a first-quantized string on a Riemannian target space X is formulated in terms of maps $x : \Sigma \to X$ with Σ a Riemann surface, that we will often choose to give the topology of a cylinder $S^1 \times \mathbb{R}$ or a torus T^2. The canonical Euclidean action, including the standard topological term, is of the form

$$\frac{1}{4\pi\alpha'} \int_\Sigma G_{\mu\nu}(x) dx^\mu \wedge *dx^\nu + \frac{i}{2\pi} \int_\Sigma x^* B + \textit{fermions} \qquad (3.1)$$

with G the Riemannian metric and B a closed two-form on X.

Here $\alpha' \in \mathbb{R}^+$ is a parameter that can be interpreted as Planck's constant. An important feature of the two-dimensional sigma model is that in the limit $\alpha' \to 0$ it reduces to the supersymmetric quantum mechanics of the previous section. This limit can be equivalently seen as a rescaling of the metric G and thereby a low-energy or a large-volume limit, $vol(X) \to \infty$. In this point-particle limit the dependence on the B-field disappears.

In the Hamiltonian formulation one describes a single string moving on a space X in terms of the loop space $\mathcal{L}X$ of maps $S^1 \to X$. Depending on the particular type of string theory that we are interested in, this first-quantization leads us to assign to the manifold X a single string superconformal field theory (SCFT) Hilbert space

$$\mathcal{Q}^1 : X \to \mathcal{H}(X), \qquad (3.2)$$

that can be formally considered to be the space of half-infinite dimensional differential forms on $\mathcal{L}X$. We will always choose in the definition of $\mathcal{H}$ Ramond or periodic boundary conditions for the fermions. These boundary conditions respect the supersymmetry algebra; other boundary conditions can be obtained by spectral flow [23].

On this Hilbert space act two natural operators: the Hamiltonian H, roughly the generalized Laplacian on $\mathcal{L}X$, and the momentum operator P that generates the canonical circle action on the loop space corresponding to rotations of the loop,

$$e^{i\theta P} : x(\sigma) \mapsto x(\sigma + \theta). \qquad (3.3)$$

In a conformal field theory the operators H and P are usually written in terms of left-moving and right-moving Virasoro generators L_0 and $\overline{L}_0$ as

$$H = L_0 + \overline{L}_0 - d/4, \qquad P = L_0 - \overline{L}_0. \qquad (3.4)$$

Here d is the *complex* dimension of X. If the manifold X is a Calabi-Yau space, the quantum field theory carries an $N = 2$ superconformal algebra with a $U(1)_L \times U(1)_R$ R-symmetry. In particular this allows us to define separate left-moving and right-moving conserved fermion numbers F_L and F_R, that up to an infinite shift (that is naturally regularized in the QFT) represent the bidegrees in terms of the Dolbeault differential forms on $\Omega^*(\mathcal{L}X)$.

The most general partition function is written as

$$Z(X;q,y,\bar{q},\bar{y}) = \text{Tr}_{\mathcal{H}}(-1)^F y^{F_L}\bar{y}^{F_R}q^{L_0-\frac{d}{8}}\bar{q}^{\bar{L}_0-\frac{d}{8}} \tag{3.5}$$

with $F = F_L + F_R$ the total fermion number. The partition function Z represents the value of the path-integral on a torus or elliptic curve, and we can write $q = e^{2\pi i\tau}, y = e^{2\pi iz}$ with τ the modulus of the elliptic curve and z a point in its Jacobian that determines the line-bundle of which the fermions are sections. The spectrum of all four operators $L_0, \bar{L}_0, F_L, F_R$ is discrete with the further conditions

$$L_0, \bar{L}_0 \geq d/8, \qquad L_0 - \bar{L}_0 \in \mathbb{Z}, \qquad F_L, F_R \in \mathbb{Z} + \tfrac{d}{2}. \tag{3.6}$$

For a general Calabi-Yau manifold it is very difficult to compute the above partition function explicitly. Basically, only exact computations have been done for orbifolds and the so-called Gepner points, which are spaces with exceptionally large quantum automorphism groups. This is not surprising, since even in the $\alpha' \to 0$ limit we would need to know the spectrum of the Laplacian, while for many Calabi-Yau spaces such as $K3$ surfaces an explicit Ricci flat metric is not even known.

Just as for the quantum mechanics case, we learn a lot by considering the supersymmetric ground states $\psi \in V \subset \mathcal{H}$ that satisfy $H\psi = 0$. In the Ramond sector the ground states are canonically in one-to-one correspondence with the cohomology classes in the Dolbeault groups,

$$\mathcal{V} \cong H^{*,*}(X). \tag{3.7}$$

In fact, these states have special values for the conserved charges. Ramond ground states always have $L_0 = \bar{L}_0 = d/8$, and for a ground state that corresponds to a cohomology class $\psi \in H^{r,s}(X)$ the fermion numbers are shifted degrees

$$F_L = r - d/2, \qquad F_R = s - d/2. \tag{3.8}$$

The shift in degrees by $d/2$ is a result from the fact that we had to 'fill up' the infinite Fermi sea. We see that there is an obvious reflection symmetry $F_{L,R} \to -F_{L,R}$ (Poincaré duality) around the middle dimensional cohomology. If we take the limit $q, \bar{q} \to 0$, the partition function reduces essentially to the Poincaré-Hodge polynomial of X

$$\begin{aligned}
h(X;y,\bar{y}) &= \lim_{q,\bar{q}\to 0} Z(X;q,\bar{q},y,\bar{y}) \\
&= \sum_{0\leq r,s\leq d} (-1)^{r+s} y^{r-\frac{d}{2}}\bar{y}^{s-\frac{d}{2}} h^{r,s}(X)
\end{aligned} \tag{3.9}$$

3.2   The elliptic genus

An interesting specialization of the sigma model partition function is the elliptic
genus of X [24], defined as

$$\chi(X;q,y) = \mathrm{Tr}_{\mathcal{H}}(-1)^F y^{F_L} q^{L_0 - \frac{d}{8}} \tag{3.10}$$

The elliptic genus is obtained as a specialization of the general partition function
for $\bar{y} = 1$. Its proper definition is

$$\chi(X;q,y) = \mathrm{Tr}_{\mathcal{H}}(-1)^F y^{F_L} q^{L_0 - \frac{d}{8}} \bar{q}^{\bar{L}_0 - \frac{d}{8}} \tag{3.11}$$

But, just as for the Witten index, because of the factor $(-1)^{F_R}$ there are no
contributions of states with $\bar{L}_0 - d/8 > 0$. Only the right-moving Ramond ground
states contribute. The genus is therefore holomorphic in q or τ. Since this fixes
$L_0 - d/8$ to be an integer, the partition function becomes a topological index,
with no dependence on the moduli of X.

Using general facts of modular invariance of conformal field theories, one
deduces that for a Calabi-Yau d-fold the elliptic genus is a weak Jacobi form [25]
of weight zero and index $d/2$. (For odd d one has to include multipliers or work
with certain finite index subgroups, see [26, 27, 28].) The ring of Jacobi forms is
finitely generated, and thus finite-dimensional for fixed index[4]. It has a Fourier
expansion of the form

$$\chi(X;q,y) = \sum_{m \geq 0,\, \ell} c(m,\ell) q^m y^\ell \tag{3.12}$$

with integer coefficients. The terminology 'weak' refers to the fact that the term
$m = 0$ is included.

The elliptic genus has beautiful mathematical properties. In contrast with
the full partition function, it does not depend on the moduli of the manifold X:
it is a (differential) topological invariant. In fact, it is a genus in the sense of
Hirzebruch — a ring-homomorphism from the complex cobordism ring $\Omega_U^*(pt)$
into the ring of weak Jacobi forms. That is, it satisfies the relations

$$\chi(X \cup X';q,y) = \chi(X;q,y) + \chi(X';q,y),$$
$$\chi(X \times X';q,y) = \chi(X;q,y) \cdot \chi(X';q,y), \tag{3.13}$$
$$\chi(X;q,y) = 0, \qquad \text{if } X = \partial Y,$$

---

[4] For example, in the case $d = 2$ it is one-dimensional and generated by the elliptic genus of
$K3$.

where the last relation is in the sense of complex bordism. The first two relations are obvious from the quantum field theory point of view; they are valid for all partition functions of sigma models. The last condition follows basically from the definition in terms of classical differential topology, more precisely in terms of Chern classes of symmetrized products of the tangent bundle, that we will give in a moment. We already noted that in the limit $q \to 0$ the genus reduces to a weighted sum over the Hodge numbers, which is essentially the Hirzebruch χ_y-genus,

$$\chi(X; 0, y) = \sum_{r,s} (-1)^{r+s} h^{r,s}(X) y^{r-\frac{d}{2}}, \tag{3.14}$$

and for $y = 1$ it equals the Witten index or Euler number of X

$$\chi(X; q, 1) = \mathrm{Tr}_{\mathcal{H}}(-1)^F = \chi(X). \tag{3.15}$$

For smooth manifolds, the elliptic genus has an equivalent definition as

$$\chi(X; q, y) = \int_X ch(E_{q,y}) td(X) \tag{3.16}$$

with the formal sum of vector bundles

$$E_{q,y} = y^{-\frac{d}{2}} \bigotimes_{n>0} \left(\Lambda_{-yq^{n-1}} T_X \otimes \Lambda_{-y^{-1}q^n} \overline{T}_X \otimes S_{q^n} T_X \otimes S_{q^n} \overline{T}_X \right), \tag{3.17}$$

where T_X denotes the holomorphic tangent bundle of X. If the bundle $E_{y,q}$ is expanded as

$$E_{q,y} = \bigoplus_{m,\ell} q^m y^\ell E_{m,\ell} \tag{3.18}$$

the coefficients $c(m,\ell)$ give the index of the Dirac operator on X twisted with the vector bundle $E_{m,\ell}$, and are therefore integers. This definition follows from the sigma model by taking the large volume limit, where curvature terms can be ignored and one essentially reduces to the free model, apart from the zero modes that give the integral over X.

3.3 Physical interpretation of the elliptic genus

Physically, the elliptic genus arises in two interesting circumstances. First, it appears as a counting function of perturbative string BPS states. (BPS states are states that form small representations of the supersymmetry algebra.) If one constrains the states of the string to be in a right-moving ground state, $i.e.$, to satisfy $\overline{L}_0 = d/8$, the states are invariant under part of the space-time supersymmetry algebra and called BPS. The generating function of such states is naturally given by the elliptic genus. Because we weight the right-movers with the chiral Witten index $(-1)^{F_R}$, only the right-moving ground states contribute.

Another physical realization is the so-called half-twisted model. Starting from a $N = 2$ superconformal sigma model, we can obtain a topological sigma model, by changing the spins of the fermionic fields. This produces two scalar nilpotent BRST operators Q_L, Q_R that can be used to define cohomological field theories. If we use both operators, or equivalently the combination $Q = Q_L + Q_R$, the resulting field theory just computes the quantum cohomology of X. This topological string theory is the appropriate framework to understand the Gromov-Witten invariants. If we ignore interactions for the moment, the free spectrum is actually that of a quantum field theory. Indeed, the gauging implemented by the BRST operator removes all string oscillations, forcing the states to be both left-moving and right-moving ground states

$$L_0\psi = \overline{L}_0\psi = 0. \tag{3.19}$$

Only the harmonic zero-modes contribute. In this way one finds one quantum field for every differential form on the space-time. This is precisely the model that we discussed in the previous section.

However, as first suggested by Witten in [29], it is also possible to do this twist only for the right-moving fields. In that case, we have to compute the cohomology of the right-moving BRST operator Q_R. This cohomology has again harmonic representatives with $\overline{L}_0 = 0$. These states coincide with the BPS states mentioned above. The half-twisted cohomology is no longer finite-dimensional, but it is graded by L_0 and F_L and the dimensions of these graded pieces are encoded in the elliptic genus. So, the half-twisted string is a proper string theory with an infinite tower of heavy states.

3.4   Second-quantized elliptic genera

We now come to the analogue of the theorem of Göttsche and Hirzebruch for the elliptic genus as it was conjectured in [30] and derived in [10].

Theorem 2 [10] — *The orbifold elliptic genus of the symmetric products $S^N X$ is given by*

$$\chi_{orb}(S_p X; q, y) = \prod_{n>0,\, m\geq 0,\, \ell} (1 - p^n q^m y^\ell)^{-c(nm,\ell)} \tag{3.20}$$

In order to prove this result, we have to compute the elliptic genus or, more generally, the string partition function for the orbifold M/G with $M = S^N X$ and $G = S_N$. The computation follows closely the computation of the orbifold Euler character that was relevant for the point-particle case.

First of all, the decomposition of the Hilbert space in superselection sectors labeled by the conjugacy class of an element $g \in G$ follows naturally. The superconformal sigma model with target space M can be considered as a quantization of the loop space $\mathcal{L}M$. If we choose as our target space an orbifold M/G, the loop space $\mathcal{L}(M/G)$ will have disconnected components of loops in M satisfying the twisted boundary condition

$$x(\sigma + 2\pi) = g \cdot x(\sigma), \qquad g \in G, \tag{3.21}$$

and these components are labeled by the conjugacy classes $[g]$. In this way, we find that the Hilbert space of any orbifold conformal field theory decomposes naturally into twisted sectors. Furthermore, in the untwisted sector we have to take the states that are invariant under G. For the twisted sectors we can only take invariance under the centralizer C_g, which is the largest subgroup that commutes with g. If $\mathcal{H}_g$ indicates the sector twisted by g, the orbifold Hilbert space has therefore the general form [31]

$$\mathcal{H}(M/G) = \bigoplus_{[g]} \mathcal{H}_g^{C_g} \tag{3.22}$$

In the point-particle limit $\alpha' \to 0$ the size of all loops shrinks to zero. For the twisted boundary condition this means that the loop gets necessarily concentrated on the fixed point set M^g and we are in fact dealing with a point-particle on M^g/C_g. In this way the string computation automatically produces the prescription for the orbifold cohomology that we discussed before. Indeed, as we stress, the quantum mechanical model of the previous section can best be viewed as a low-energy limit of the string theory.

In the case of the symmetric product $S^N X$, the orbifold superselection sectors correspond to partitions $\{N_n\}$ of N. Furthermore, we have seen that for a given partition the fixed point locus is simply the product

$$\prod_n S^{N_n} X^{(n)} \tag{3.23}$$

Here we introduce the notation $X^{(n)}$, to indicate a copy of X obtained as the diagonal in X^N where n points coincide. In the case of point-particles this distinction was not very important but for strings it is absolutely crucial.

The intuition is best conveyed with the aid of *Fig. 1* where we depicted a generic twisted sector of the orbifold sigma model. The crucial point is that such a configuration can be interpreted as describing long strings[5] whose number can be smaller than N. Indeed, as we clearly see, a twisted boundary condition containing a elementary cycle of length n gives rise to a single string of 'length' n built out of n 'string bits.' If the cycle permutes the coordinates $(x_1, \ldots, x_n) \in X^n$ as

$$x_k(\sigma + 2\pi) = x_{k+1}(\sigma), \qquad k \in (1, \ldots, n), \tag{3.24}$$

we can construct a new loop $x(\sigma)$ by gluing the n strings $x_1(\sigma), \ldots, x_n(\sigma)$ together:

---

[5] The physical significance of this picture was developed in among others [32, 33, 34] and made precise in [10].

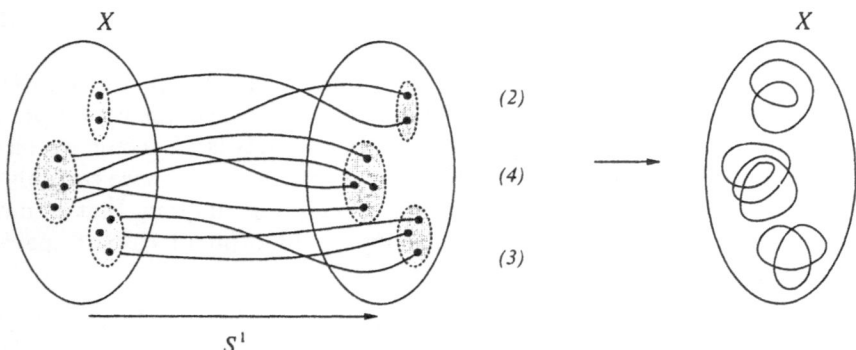

Fig. 1: *A twisted sector of a sigma model on $S^N X$ can describe less than N strings. (Here $N = 9$ and the sector contains three 'long strings.')*

$$x(\sigma) = x_k(\sigma') \quad \text{if} \quad \sigma = \tfrac{1}{n}\big(2\pi(k-1) + \sigma'\big), \ \sigma' \in [0, 2\pi]. \tag{3.25}$$

If the twist element is the cycle $(N) \in S_N$, such a configuration describes one single long string of length N, instead of the N short strings that we would expect.

In this fashion we obtain from a cyclic twist (n) one single copy of the loop space $\mathcal{L}X$ that we denote as $\mathcal{L}X^{(n)}$. We use the notation $\mathcal{H}^{(n)}$ for its quantization. The twisted loop space $\mathcal{L}X^{(n)}$ is distinguished from the untwisted loop space $\mathcal{L}X$ in that the canonical circle action is differently normalized. We now have

$$e^{i\theta P} : \ x(\sigma) \to x(\sigma + \theta/n). \tag{3.26}$$

So we find that only for $\theta = 2\pi n$ do we have a full rotation of the loop. This is obvious from the twisted boundary condition (3.24). It seems to imply that the eigenvalues for the operator $P = L_0 - \overline{L}_0$ in this sector are quantized in units of $1/n$. Together with the fact that in the elliptic genus only states with $\overline{L}_0 = 0$ contribute, this would suggest that the contribution of the sector $\mathcal{H}^{(n)}$ to the elliptic genus is[6]

$$\chi(\mathcal{H}^{(n)}; q, y) \overset{?}{=} \sum_{m,\ell} c(m, \ell) q^{m/n} y^{\ell}. \tag{3.27}$$

However, we must remember that the centralizer of a cycle of length n contains a factor $\mathbb{Z}_n$. This last factor did not play a role in the point-particle case, but here it does act non-trivially. In fact, it is precisely generated by $e^{2\pi i P}$. The orbifold definition includes a prescription to take the states that are invariant under the action of the centralizer. So only the states with integer eigenvalues of P survive. In this way only the states with m congruent to 0 modulo n survive and we obtain an integer q-expansion,

---

[6] We use the more general notation $\chi(\mathcal{H}; q, y) = \mathrm{Tr}_{\mathcal{H}}(-1)^F q^{L_0 - \frac{d}{8}} y^{F_L}$ for any Hilbert space $\mathcal{H}$.

$$\chi(\mathcal{H}^{(n)}; y, q) = \sum_{m,\ell} c(nm, \ell) q^m y^\ell.$$ (3.28)

We now again assemble the various components to finish the proof of (3.20) (for more details see [10]).

$$\sum_{N \geq 0} p^N \chi_{orb}(S^N X; q, y) = \sum_{N \geq 0} p^N \sum_{\substack{N_n \\ \sum n N_n = N}} \prod_{n > 0} \chi(S^{N_n} \mathcal{H}^{(n)}; q, y)$$

$$= \prod_{n > 0} \sum_{N \geq 0} p^{nN} \chi(S^N \mathcal{H}^{(n)}; q, y)$$

$$= \prod_{n > 0, \, m, \, \ell} (1 - p^n q^m y^\ell)^{-c(nm, \ell)}$$ (3.29)

The infinite product formula has strong associations to automorphic forms and denominator formulas of generalized Kac-Moody algebras [35] and string one-loop amplitudes [36]; we will return to this.

3.5 General partition function

It is not difficult to repeat the above manipulations in symmetric algebra for the full partition function. In fact, we can write a general formula for the second-quantized string Fock space, similar as we did for the point-particle case in (2.35). This Fock space is again of the form

$$\mathcal{F}_p = \bigotimes_{n > 0} S_{p^n} \mathcal{H}^{(n)}.$$ (3.30)

Here $\mathcal{H}^{(n)}$ is the Hilbert space obtained by quantizing a single string that is wound n times. It is isomorphic to the subspace of the single string Hilbert space $\mathcal{H} = \mathcal{H}^{(1)}$ with

$$L_0 - \overline{L}_0 = 0 \pmod{n}.$$ (3.31)

The action of the operators L_0 and $\overline{L}_0$ on $\mathcal{H}^{(n)}$ are then rescaled by a factor $1/n$ compared with the action on $\mathcal{H}$

$$L_0^{(n)} = L_0^{(1)}/n, \qquad \overline{L}_0^{(n)} = \overline{L}_0^{(1)}/n.$$ (3.32)

As we explained already, this rescaling is due to the fact that the string has now length $2\pi n$ instead of 2π. Even though the world-sheet Hamiltonians $L_0^{(n)}, \overline{L}_0^{(n)}$ have fractional spectra compared to the single string Hamiltonians, the momentum operator still has an integer spectrum,

$$L_0^{(n)} - \overline{L}_0^{(n)} = 0 \ (\mathrm{mod} \ 1), \tag{3.33}$$

due to the restriction (3.31) that is implemented by the orbifold $\mathbb{Z}_n$ projection.

It is interesting to reconsider the ground states of $\mathcal{H}^{(n)}$, in particular their $U(1)_L \times U(1)_R$ charges, since this will teach us something about the orbifold cohomology of $S^N X$. A ground state $\psi^{(n)} \in \mathcal{V}^{(n)}$ that corresponds to a cohomology class $\psi \in H^{r,s}(X)$ still has fermion charges F_L, F_R given by

$$F_L = r - d/2, \qquad F_R = s - d/2. \tag{3.34}$$

Making the string longer does not affect the $U(1)$ current algebra. However, since these states now appear as ground states of a conformal field theory with target space $S^N X$, which is of complex dimension $n \cdot d$, these fermion numbers have a different topological interpretation. The corresponding degrees $r^{(n)}, s^{(n)}$ of the same state now considered as a differential form in the orbifold cohomology of $S^n X \subset S^N X$ are therefore shifted as

$$r^{(n)} = r + (n-1)d/2, \qquad s^{(n)} = s + (n-1)d/2. \tag{3.35}$$

That is, we have

$$\mathcal{V}^{(n)} \cong H^{*-(n-1)\frac{d}{2}, *-(n-1)\frac{d}{2}}(X). \tag{3.36}$$

In the quantum mechanics limit, the twisted loops that give rise to the contribution $\mathcal{H}^{(n)}$ in the Fock space become point-like and produce another copy $X^{(n)}$ of the fixed point set X. However, this copy of X is the small diagonal in X^n. We see that this gives another copy of $H^*(X)$ however now shifted in degree. In the full Fock space we have an infinite number of copies, shifted by positive multiples of $(d/2, d/2)$.

We can encode all this in the generating function of Hodge numbers (3.9) as

$$h_{orb}(S_p X; y, \overline{y}) = \prod_{n>0, r, s} (1 - p^n y^r \overline{y}^s)^{-(-1)^{r+s} h^{r,s}(X)} \tag{3.37}$$

For the full partition function we can write a similar expression

$$Z_{orb}(S_p X; q, \overline{q}, y, \overline{y}) = \prod_{n>0} \prod_{\substack{h, \overline{h}, r, s \\ h - \overline{h} = 0 \ (\mathrm{mod} \ n)}} \left(1 - p^n q^h \overline{q}^{\overline{h}} y^r \overline{y}^s\right)^{-c(h, \overline{h}, r, s)} \tag{3.38}$$

where

$$Z(X; q, \overline{q}, y, \overline{y}) = \sum_{h, \overline{h}, r, s} c(h, \overline{h}, r, s) q^h \overline{q}^{\overline{h}} y^r \overline{y}^s \tag{3.39}$$

is the single string partition function.

4 Light-Cone Quantization of Quantum Field Theories

We now turn to the physical interpretation of the above results. Usually quantum field theories are quantized by splitting, at least locally, a Lorentzian space-time in the form $\mathbb{R} \times \Sigma$ where $\mathbb{R}$ represents the time direction and Σ is a space-like Cauchy manifold. Classically, one specifies initial data on Σ which then deterministically evolve through some set of differential equations in time. In recent developments it has proven useful to use for the time direction a null direction. This complicates of course the initial value problem, but has some other advantages. One can try to see this as a limiting case where one uses Lorentz transformations to boost the time-like direction to an almost null direction [37].

4.1 The two-dimensional free scalar field revisited

Before we turn to the interpretation of our results on second quantization and symmetric products in terms of quantum field theory and quantum string theory, let us first revisit one of the simplest examples of a QFT and compute the partition function of a two-dimensional free scalar field. In fact, let us be slightly more general and consider a finite number c of such scalar fields labeled by a c-dimensional real vector space V. (One could easily take this vector space to be graded, but for simplicity we assume it to be even.)

Quantization of this model usually proceeds as follows: one chooses a two-dimensional space-time with the topology of a cylinder $\mathbb{R} \times S^1$ and with coordinates (x^0, x^1). One then introduces the light-cone variables $x^{\pm} = x^0 \pm x^1$. A classical solution of the equation of motion $\Delta\phi = 0$ is decomposed as

$$\phi(x^+, x^-) = q + px^0 + \phi_L(x^-) + \phi_R(x^+), \tag{4.1}$$

where the zero-mode contribution (describing a point particle on V with coordinate q and conjugate momentum p) is isolated from the left-moving and right-moving oscillations $\phi_L(x^-)$ and $\phi_R(x^+)$. The non-zero modes have a Fourier expansion

$$\phi_L(x^-) = \sum_{n \neq 0} \tfrac{1}{n} \alpha_n e^{inx^-} \tag{4.2}$$

with a similar expression for $\phi_R(x^+)$.

In canonical quantization the Fourier coefficients α_n are replaced by creation and annihilation operators with commutation relations $[\alpha_n, \alpha_m] = n\delta_{n+m}$. This Heisenberg algebra is realized on a Fock space $\mathcal{F}$ built on a vacuum state $|0\rangle$ satisfying $\alpha_n|0\rangle = 0$ for $n > 0$. This Fock space can be written in terms of symmetric products as

$$\mathcal{F}_p = \bigotimes_{n>0} S_{p^n} V^{(n)} = S^* \left(\bigoplus_{n>0} p^n V^{(n)} \right), \tag{4.3}$$

where $V^{(n)}$ is a copy of V with the property that the (chiral) oscillation number operator N has eigenvalue n on $V^{(n)}$. The p-expansion keeps track of the N-gradation. As a quantum operator N is defined as

$$N = \sum_{n>0} \alpha_{-n}\alpha_n. \tag{4.4}$$

The chiral partition function is now written as a character of this module (with $p = e^{2\pi i \tau}$, τ in the upper half-plane $\mathbb{H}$)

$$\chi(p) = \dim \mathcal{F}_p = \mathrm{Tr}_{\mathcal{F}}p^N = \prod_{n>0}(1-p^n)^{-c} \tag{4.5}$$

This character is almost a modular form of weight $-c/2$. One way to see this, is by considering the partition function of the full Hilbert space $\mathcal{H}$, which is modular invariant. $\mathcal{H}$ is obtained by combining the left-moving oscillators with the right-moving oscillators and adding the zero-mode contribution

$$\mathcal{H} = L^2(V) \otimes \mathcal{F} \otimes \overline{\mathcal{F}} \tag{4.6}$$

The total chiral Hamiltonians can be written as

$$L_0 = -\tfrac{1}{2}\Delta + N, \qquad \overline{L}_0 = -\tfrac{1}{2}\Delta + \overline{N}. \tag{4.7}$$

The full partition function is then evaluated as

$$Z(p,\overline{p}) = \mathrm{Tr}_{\mathcal{H}}p^{L_0-c/24}\overline{p}^{\overline{L}_0-c/24} = \left(\sqrt{\mathrm{Im}\,\tau}\,|\eta(p)|^2\right)^{-c} \tag{4.8}$$

with $\eta(p)$ the Weierstrass eta-function

$$\eta(p) = p^{\frac{1}{24}}\prod_{n>0}(1-p^n). \tag{4.9}$$

So the proper modular object is given by

$$\mathrm{Tr}_{\mathcal{F}}p^{N-c/24} = \eta(p)^{-c} \tag{4.10}$$

which is a modular form of $SL(2,\mathbb{Z})$ of weight $-c/2$ (with multipliers if $c \neq 0$ (mod 24).) The extra factor $p^{-c/24}$ is interpreted as a regularized infinite sum of zero-point energies that appear in canonical quantization.

In the Lagrangian formalism the same result is given in terms of a ζ-function regularized determinant of c scalar fields

$$Z = (\sqrt{\mathrm{Im}\,\tau}/\det{}'\Delta)^{c/2} \tag{4.11}$$

with Δ the Laplacian on the torus T^2 and the prime indicates omission of the zero-mode. This determinant can be computed in a first-quantized form as a one-loop integral

$$\log Z = -\tfrac{1}{2}c \log \det \Delta = -\tfrac{1}{2}c \mathrm{Tr} \log \Delta = c \int_0^\infty \frac{dt}{t}\, \mathrm{Tr}_{\mathcal{H}}\, e^{-tH} \qquad (4.12)$$

where $\mathcal{H} = L^2(T^2)$ is now the quantum mechanical Hilbert space of a single particle moving on T^2. Here the RHS is defined by cutting off the integral at $t = \epsilon$ and subtracting the ϵ-dependent (but τ-independent) term.

Let us mention a few aspects of these results that we will try to generalize when we consider strings instead of quantum fields in the next section.

 (i) The quantum field theory partition function factorizes in left-moving and right-moving contributions that are holomorphic functions of the modulus p.

 (ii) The holomorphic contributions are modular forms of weight $-c/2$ under $SL(2,\mathbb{Z})$ if a particular correction (here $p^{-c/24}$) is added.

 (iii) The full partition function is modular invariant because the zero-mode contribution adds a non-holomorphic factor $(\mathrm{Im}\,\tau)^{-c/2}$.

 (iv) The holomorphic contributions are characters of an infinite-dimensional Kac-Moody algebra, in fact in this simple case just the Heisenberg algebra generated by the operators α_n.

 (v) The modularity of the characters, *i.e.*, the transformation properties under the modular group $SL(2,\mathbb{Z})$ is 'explained' by the relation to a partition function of a quantum field on a two-torus T^2 with modulus τ and automorphism group $SL(2,\mathbb{Z})$.

4.2   Discrete light-cone quantization

In light-cone quantization one works on $\mathbb{R}^{1,1}$ in terms of the light-cone coordinates $x^\pm$ with metric

$$ds^2 = 2dx^+ dx^-, \qquad (4.13)$$

but now chooses the null direction x^+ as the time coordinate. We will write the conjugate momenta as

$$p^+ = p_- = -i\frac{\partial}{\partial x^-}, \qquad p^- = p_+ = -i\frac{\partial}{\partial x^+}. \qquad (4.14)$$

In the usual euclidean formulation of two-dimensional CFT we have $p^+ = L_0$, $p^- = \overline{L}_0$. In this setup a free particle has an eigentime given by $x^+ = p^+ t$. The light-cone Hamiltonian describes evolution in the 'light-cone time' x^+ and so is given by

$$H_{lc} = p^- \tag{4.15}$$

An initial state is specified by the x^--dependence for fixed x^+.

The so-called discrete light-cone quantization (DLCQ) further assumes that the null direction x^- is compact of radius R

$$x^- \sim x^- + 2\pi R. \tag{4.16}$$

(The specific value of R is not very important since it can of course be rescaled by a Lorentz boost. We will therefore often put it to one, $R = 1$.) We denote the Lorentzian manifold so obtained as $(\mathbb{R} \times S^1)^{1,1}$. The periodic identification of x^- makes the spectrum of the conjugate momentum p^+ discrete

$$p^+ \in N/R, \qquad N \in \mathbb{Z}. \tag{4.17}$$

Now we have for fixed x^+ a decomposition of the scalar field as

$$\phi(x^-) = \sum_{n \neq 0} \tfrac{1}{n} \alpha_n e^{inx^-} \tag{4.18}$$

(Clearly, this quantization scheme is incomplete, since we are omitting the zero-modes with $p^+ = 0$. We will only obtain the left-moving sector of the theory. We will return to this point.) Since the classical equation for a free field reads

$$\partial_+ \partial_- \phi = 0, \tag{4.19}$$

the field $\phi(x^-)$ will have no x^+-dependence and therefore the light-cone energy of its modes will vanish, $p^- = 0$. If we have c of these free scalar fields $\phi(x) \in V$, quantization will result in the same chiral Fock space that we considered in the canonical quantization

$$\mathcal{F}_p = \bigotimes_{n>0} S_{p^n} V^{(n)} \tag{4.20}$$

and the light-cone partition function is given by same infinite product

$$\mathrm{Tr}_{\mathcal{F}}\, p^{P^+} = \prod_{n>0} (1 - p^n)^{-c} \tag{4.21}$$

Note that the eigenvalues of the longitudinal momentum $P^+ = n$ are always positive. This is due to the fact that the oscillation numbers α_n form a Heisenberg algebra.

We recognize in these formulas our computations of the Euler number of the symmetric products of a space X with $\chi(X) = c$. To explain this relation we now consider the light-cone quantization of field theories in higher dimensions.

4.3 Higher-dimensional scalar fields in DLCQ

Things become a bit more interesting if we consider a free scalar field on a more general space-time of the form

$$M^{1,d+1} = (\mathbb{R} \times S^1)^{1,1} \times X^d, \tag{4.22}$$

X compact Riemannian and with light-cone coordinates (x^+, x^-, x). We adopt light-cone quantization and consider as initial data the field configuration on $x^+ = constant$. The light-cone Hamiltonian p^- again describes the evolution in x^+. It will be convenient to perform a Fourier transformation in the light-cone coordinate x^- and consider a basis of field configurations of the form

$$\phi(x^+, x^-, x) = e^{i(p^- x^+ + p^+ x^-)} \phi_m(x) \tag{4.23}$$

with $p^+ = n$ (we put $R = 1$) and $\phi_m(x)$ an eigenstate of the transverse Hamiltonian $H = -\frac{1}{2}\Delta^{(X)}$,

$$H\phi_m = h_m \phi_m. \tag{4.24}$$

The equation of motion on the space-time M, $\Delta^{(M)}\phi = 0$, then gives the so-called mass-shell relation

$$p^- = -\frac{1}{2p^+}\Delta^{(X)} = \frac{1}{n}h_m. \tag{4.25}$$

Here we see an interesting phenomenon. The light-cone energy is given by a non-relativistic expression of the form $p^2/2m$, where p is the transversal momentum and the 'mass' m is given by the longitudinal momentum $p^+ = n$. (On a curved manifold p^2 is replaced by the eigenvalues h_m of the Laplacian.) The appearance of this non-relativistic expression has its geometric explanation in the fact that the stabilizer group of a null-direction in $\mathbb{R}^{1,n+1}$ is the Galilean group of $\mathbb{R}^n$. The formula implies that for a particle with $p^+ = n$, the light-cone energy p^- is rescaled by a factor of $1/n$.

Note that quite generally in light-cone quantization the symmetries of the underlying space-time manifold are not all manifest. If we work with the Minkowski space-time $\mathbb{R}^{1,n+1}$ the Lorentz group $SO(1, n+1)$ is partly non-linearly realized. For interacting QFT's the proof of Lorentz invariance of a light-cone formulation is highly non-trivial. In DLCQ the Lorentz-invariance is only expected in the limit $R \to \infty$. Since the value of R can be rescaled by a Lorentz boost, this limit is equivalent to the large N limit, $N \to \infty$. Again, for interacting theories the appearance of Lorentz-invariance in this limit is not obvious.

Upon quantization we obtain in the present case a Fock space that is of the form

$$\mathcal{F}_p = \bigotimes_{n>0} \bigoplus_{N \geq 0} p^{nN} S^N \mathcal{H}^{(n)} \tag{4.26}$$

with $\mathcal{H}^{(n)} = L^2(X)$ with $p^- = H^{(n)} = H/n$ the rescaled QM Hamiltonian. Now the light-cone energies p^- will not typically vanish, and the full partition function is given by a two-variable function,

$$Z(X,p,q) = \text{Tr}_{\mathcal{F}} p^{P^+} q^{P^-}, \tag{4.27}$$

with P^+, P^- the total light-cone momentum operators (with eigenvalues p^+, p^-).

From the above description it should have become clear that this partition function can be completely identified with the quantum mechanics on the symmetric product space $S_p X = \coprod_N p^N S^N X$ that we discussed in such detail in section 2. We therefore obtain:

Theorem 3 — *The discrete light-cone quantization of a free scalar field on the space-time $M = (\mathbb{R} \times S^1)^{1,1} \times X$ with total longitudinal momentum $p^+ = N$ is given by the quantum mechanics on the orbifold symmetric product $S^N X$,*

$$\mathcal{H}^{QFT}(X) = \mathcal{H}_{orb}^{QM}(SX). \tag{4.28}$$

Furthermore, the light-cone Hamiltonian p^- is identified with the non-relativistic quantum mechanics Hamiltonian H.

4.4   The supersymmetric generalization

It is easy to extend this construction to a physical system that describes the supersymmetric quantum mechanics on SX. In that case we want to have arbitrary differential forms on X, so our fundamental fields will be free k-forms $\phi^k \in \Omega^k(M)$ with $0 \leq k \leq d$ on the space-time $M = (\mathbb{R} \times S^1)^{1,1} \times X$. These fields have a quadratic action (with fermionic statistics if k is odd)

$$\int_Y \tfrac{1}{2} d\phi \wedge *d\phi, \qquad \phi = \sum_k \phi^k \in \Omega^*(Y). \tag{4.29}$$

This gives as equation of motion the Maxwell equation

$$d^* d\phi = 0. \tag{4.30}$$

This Lagrangian is invariant under the gauge symmetry

$$\phi^k \to \phi^k + d\lambda^{k-1}, \qquad \lambda \in \Omega^{k-1}(M), \tag{4.31}$$

giving ϕ^k the interpretation of a generalized k-form connection with curvature $d\phi^k$. This gauge symmetry can be fixed by requiring

$$\iota_{\partial/\partial x^+} \phi = 0, \tag{4.32}$$

a condition that we write as $\phi_+ = 0$. With this gauge condition the equation of motion can be used to eliminate the component ϕ_- in terms of the transversal components $\phi \in \Omega^*(X)$, leaving only the form on the transverse space X as physical. All of this is well-known from the description of the Ramond-Ramond (RR) fields of the light-cone type II superstring (or supergravity).

With this gauge fixing we naturally reduce the second-quantized light-cone description to supersymmetric quantum mechanics on SX. We therefore find exactly the field theoretic description of our SQM model of section 2. It describes the multi-form abelian gauge field theory on $M = (\mathbb{R} \times S^1)^{1,1} \times X$ in DLCQ, in particular we have

$$\mathcal{H}^{QFT}(X) = \mathcal{H}^{SQM}_{orb}(S_p X) = \bigotimes_n S_{p^n} \Omega^*(X)^{(n)} \qquad (4.33)$$

where powers of p keep track of the longitudinal momentum p^+.

Particularly interesting is the zero-energy $p^- = 0$ sector $\mathcal{V}^{QFT} \subset \mathcal{H}^{QFT}$. Since p^- is identified with the SQM Hamiltonian, these states correspond to the ground states of the supersymmetric quantum mechanics and we have

$$\mathcal{V}^{QFT} = \bigotimes_{n>0} S_{p^n} H^*(X) \qquad (4.34)$$

and the partition function of this zero p^- sector reproduces exactly the orbifold Euler character

$$\mathrm{Tr}_{\mathcal{V}_{QFT}} (-1)^F p^{P^+} = \chi_{orb}(S_p X). \qquad (4.35)$$

The modular properties are now explained along the lines of section 1. This partition function can be computed in a Lagrangian formulation by considering the compact space-time $T^2 \times X$. The explicit T^2 factor explains the occurrence of $SL(2,\mathbb{Z})$.

The modular properties are particularly nice if we choose as our manifold X a $K3$ surface with $\chi(X) = 24$. We then almost have a modular object without multipliers,

$$\chi_{orb}(S_p X) = \frac{p}{\Delta(p)} \qquad (4.36)$$

with $\Delta(p) = \eta^{24}(p)$ the discriminant, a cusp form of weight 12 for $SL(2,\mathbb{Z})$. The correction $p^{-\chi/24}$ has again an interpretation as the regularized sum of zero-point energies. (Each boson contributes $-1/24$, each fermion $+1/24$.)

5 Light-Cone Quantization of String Theories

It is now straightforward to generalize all this to string theory along the lines

$$\text{quantum mechanics on } SX \quad \rightarrow \quad \text{quantum field theory on } X$$
$$\text{2-d conformal field theory on } SX \quad \rightarrow \quad \text{quantum string theory on } X$$

This generalization is particularly interesting since a good Lorentz-invariant description of non-perturbative second-quantized quantum string theory is absent. So we can gain something by studying the reformulation in terms of sigma models on symmetric products.

It is not difficult to give a string theory interpretation of our results in section 3 on sigma models on symmetric product spaces. Clearly we want to identify the DLCQ of string theory on $(\mathbb{R} \times S^1)^{1,1} \times X$ with the SCFT on SX. An obvious question is which type of string theory are we discussing. Indeed, the number of consistent interacting closed string theories is highly restricted: the obvious candidates are

(i) Type II and heterotic strings in 10 dimensions.

(ii) Topological strings in all even dimensions.

(iii) Non-abelian strings in 6 dimension.

Here the last example only recently emerged, and we will return to it in section 7. We will start with the Type II string.

5.1   The IIA superstring in light-cone gauge

The physical states of the ten-dimensional type II superstring are most conveniently described in the light-cone Green-Schwarz formalism. We usually think about the superstring in terms of maps of a Riemann surface Σ into flat spacetime $\mathbb{R}^{1,9}$. But in light-cone gauge we make a decomposition $\mathbb{R}^{1,1} \times \mathbb{R}^8$ with corresponding local coordinates (x^+, x^-, x^i). The physical degrees of freedom are then completely encoded in the transverse map

$$x : \ \Sigma \rightarrow \mathbb{R}^8. \tag{5.1}$$

The model has 16 supercharges (8 left-moving and 8 right-moving) and carries a $Spin(8)$ R-symmetry.

More precisely, apart from the bosonic field x, we also have fermionic fields that are defined for a general 8-dimensional transverse space X as follows. Let $S^{\pm}$ denote the two inequivalent 8-dimensional spinor representations of $Spin(8)$. We use the same notation to indicate the corresponding spinor bundles of X. Let V denote the vector representation of $Spin(8)$ and let TX be the associated tangent bundle. In this notation we have

$$\partial x \in \Gamma(K_\Sigma \otimes x^* TX), \qquad \bar{\partial} x \in \Gamma(\overline{K}_\Sigma \otimes x^* TX). \tag{5.2}$$

Now the left-moving and right-moving fermions $\theta, \bar{\theta}$ are sections of

$$\theta \in \Gamma(K_\Sigma^{1/2} \otimes x^* S^+), \qquad \bar{\theta} \in \Gamma(\overline{K}_\Sigma^{1/2} \otimes x^* S^\pm) \tag{5.3}$$

The choice of spin structure on Σ is always Ramond or periodic. The different choices of $Spin(8)$ representations for the right-moving fermion $\bar{\theta}$ (S^+ or S^-) give the distinction between the IIA and IIB string. We will work with the IIA string for which the representation of $\bar{\theta}$ is chosen to be the conjugate spinor S^-, but the IIB string follows the same pattern.

With these fields the action of the first-quantized sigma model is simply the following free CFT

$$S = \int d^2\sigma \left(\tfrac{1}{2}\partial x^i \bar{\partial} x^i + \theta^a \bar{\partial} \theta^a + \bar{\theta}^{\dot{a}} \partial \bar{\theta}^{\dot{a}} \right). \tag{5.4}$$

This model has a Hilbert space that is of the form

$$\mathcal{H} = L^2(\mathbb{R}^8) \otimes \mathcal{V} \otimes \mathcal{F} \otimes \overline{\mathcal{F}}. \tag{5.5}$$

We recognize familiar components: the bosonic zero-mode space $L^2(\mathbb{R}^8)$ describes the quantum mechanics of the center of mass $\oint x^i$ of the string. The fermionic zero-modes $\oint \theta^a$, $\oint \bar{\theta}^{\dot{a}}$ give rise to the 16×16 dimensional vector space of ground states

$$\mathcal{V} \cong (V \oplus S^-) \otimes (V \oplus S^+) \tag{5.6}$$

where the spinor representations should be considered odd. This space forms a representation of the Clifford algebra $\mathrm{Cliff}(S^+) \otimes \mathrm{Cliff}(S^-)$ generated by the fermion zero mode

$$\Gamma^a = \int \frac{d\sigma}{2\pi} \, \theta^a(\sigma), \qquad \Gamma^{\dot{a}} = \int \frac{d\sigma}{2\pi} \, \bar{\theta}^{\dot{a}}(\sigma). \tag{5.7}$$

Using the triality $S^+ \to V \to S^- \to S^+$ of $Spin(8)$ this maps to the usual Clifford representation of $\mathrm{Cliff}(V)$ on $S^+ \oplus S^-$. Finally the Fock space $\mathcal{F}$ of non-zero-modes is given by

$$\mathcal{F}_q = \bigotimes_{n>0} \left(\wedge_{q^n} S^- \otimes S_{q^n} V \right) \tag{5.8}$$

with a similar expression for $\overline{\mathcal{F}}$ with S^- replaced by S^+.

In this light-cone gauge the coordinate x^+ is given by

$$x^+(\sigma,\tau) = p^+ \tau \tag{5.9}$$

for fixed longitudinal momentum $p^+ > 0$, whereas x^- is determined by the constraints

$$\partial x^- = \frac{1}{p^+}(\partial x)^2, \qquad \overline{\partial} x^- = \frac{1}{p^+}(\overline{\partial} x)^2. \tag{5.10}$$

The Hilbert space of physical states of a single string with longitudinal momentum p^+ is given by the CFT Hilbert space $\mathcal{H}$ restricted to states with zero world-sheet momentum, the level-matching condition

$$P = L_0 - \overline{L}_0 = 0. \tag{5.11}$$

The light-cone energy p^- is then determined as by the mass-shell relation

$$p^- = \frac{1}{p^+}(L_0 + \overline{L}_0) = \frac{1}{p^+}H. \tag{5.12}$$

One can also consider DLCQ with the null coordinate x^- periodically identified with radius R. This induces two effects. First, the momentum p^+ is quantized as $p^+ = n/R$, $n \in \mathbb{Z}_{>0}$. Second, it allows the string to be wrapped around the compact null direction giving it a non-trivial winding number

$$w^- = \int_{S^1} dx^- = 2\pi m R, \qquad m \in \mathbb{Z}. \tag{5.13}$$

However, using the constraints (5.10) we find that

$$w^- = \frac{2\pi}{p^+}(L_0 - \overline{L}_0) = \frac{2\pi R}{n}(L_0 - \overline{L}_0). \tag{5.14}$$

So, in order for m to be an integer, we see that the CFT Hilbert space must now be restricted to the space $\mathcal{H}^{(n)}$ consisting of all states that satisfy the modified level-matching condition

$$P = L_0 - \overline{L}_0 = 0 \pmod{n}. \tag{5.15}$$

This is exactly the definition of the Hilbert space $\mathcal{H}^{(n)}$ in section 3.5. (In a similar spirit the uncompactified model has Hilbert space $\mathcal{H}^{(\infty)}$.) This motivates us to describe the second-quantized Type IIA string in terms of a SCFT on the orbifold

$$S^N \mathbb{R}^8 = \mathbb{R}^{8N}/S_N. \tag{5.16}$$

Indeed in this correspondence we have:

$$p^+ = N,$$
$$p^- = H = L_0 + \overline{L}_0, \tag{5.17}$$
$$w^- = P = L_0 - \overline{L}_0. \tag{5.18}$$

This gives the following form for the second-quantized Fock space

$$\mathcal{F}_p = \bigotimes_{n>0} S_{p^n} \mathcal{H}^{(n)} \tag{5.19}$$

where p keeps again track of p^+. This is both the Hilbert space of the free string theory and of the orbifold sigma model on $S_p \mathbb{R}^8$. So we can identify their partition functions

$$Z^{string}(\mathbb{R}^8; p,q,\bar{q}) = Z^{SCFT}(S_p \mathbb{R}^8; q,\bar{q}), \tag{5.20}$$

with

$$Z^{string}(\mathbb{R}^8; p,q,\bar{q}) = \mathrm{Tr}_{\mathcal{F}} p^{P^+} q^{P^- + W^-} \bar{q}^{P^- - W^-},$$

$$Z^{SCFT}(S_p \mathbb{R}^8; q,\bar{q}) = \sum_{N \geq 0} p^N \mathrm{Tr}_{\mathcal{H}(S^N \mathbb{R}^8)} q^{L_0 - N/2} \bar{q}^{\bar{L}_0 - N/2}, \tag{5.21}$$

(Here we used that the central charge is $12N$.)

Note that this sigma model is not precisely of the form as we discussed in section 3. The world-sheet fermions now transform as spinors instead of vectors of $Spin(8)$. The modifications that one has to make are however completely straight-forward. In particular, the $U(1)$'s whose quantum numbers gave the world-sheet fermion number $F_{L,R}$ only emerge if we break $Spin(8)$ to $SU(4) \times U(1) \cong Spin(6) \times Spin(2)$.

This issue is directly related to compactifications. Can we consider for the transversal space instead of $\mathbb{R}^8$ a compact Calabi-Yau four-fold X and make contact with our computations of the elliptic genus of SX? In the non-linear sigma models of section 3 the fermion fields were always assumed to take values in the (pull-back of the) tangent bundle to the target space X, whereas in the Green-Schwarz string they are sections of spinor bundles. Note however that on a Calabi-Yau four-fold we have a reduction of the structure group $SO(8)$ to $SU(4)$. Under this reduction we have the following well-known decomposition of the three 8-dimensional representations of $Spin(8)$ in terms of the representations of $SU(4) \times U(1)$ (up to triality)

$$
\begin{aligned}
V &\to \mathbf{4}_{1/2} \oplus \overline{\mathbf{4}}_{-1/2} \\
S^+ &\to \mathbf{4}_{-1/2} \oplus \overline{\mathbf{4}}_{1/2}, \\
S^- &\to \mathbf{1}_1 \oplus \mathbf{6}_0 \oplus \mathbf{1}_{-1}.
\end{aligned} \tag{5.22}
$$

So we see that, as far as the $SU(4)$ symmetry is concerned, if we only use the S^+ representation, we could just as well have worked with the standard $N = 2$ SCFT sigma-model, since this spinor bundle is isomorphic to the tangent bundle.

This remark picks out naturally the Type IIB string, whose world-sheet fermions carry only one $Spin(8)$ chirality that we can choose to be S^+. So in this formulation only the type IIB light-cone model allows for a full compactifications on a Calabi-Yau four-fold. This fact is actually well-known. The IIA string acquires an anomaly $\chi(X)/24$ that has to be cancelled by including some net number of strings [38, 39]. For the Type IIB string this translates under a T-duality to a net momentum in the vacuum; we will see this fact again in a moment. We can also work with only left-moving BPS strings that are related to the elliptic genus of the SCFT. In that case it does not matter if we choose the IIA or IIB strings.

5.2 Elliptic genera and automorphic forms

If we just want to discuss free strings, without interactions, say in light-cone gauge without insisting on Lorentz-invariance, there are many more possible strings than the ten-dimensional superstring. In particular we can consider a string whose transverse degrees of freedom are described by the $N = 2$ supersymmetric sigma model on the Calabi-Yau space X. This string has as its low-energy, massless spectrum the field theory that consists of all k-form gauge fields, that we discussed in section 4.4. This is by the way exactly the field content of the topological string, that can be defined in any (even) dimension, but which only has non-vanishing interactions without gravitational descendents in space-time dimension 6. So this critical case corresponds to choosing a transversal four-fold or complex surface X. If X has to be compact that restricts us to T^4 or $K3$. We will return to this topic in section 7.

Therefore another class of free string theories to be considered in the light-cone formulation are the 'untwisted' versions of the topological string, where we do not impose the usual BRST cohomology $Q_L = Q_R = 0$ that reduces the string to its massless fields. In fact, another interesting case is the half-twisted string (see section 3.3) in which we only impose $Q_R = 0$. For that model we expect to make contact with the elliptic genus.

Indeed, in that case there is a straightforward explanation of the automorphic properties of the elliptic genus of the symmetric product. We recall the main formula (3.20) of Theorem 2, that we now interpret as a partition function of second-quantized BPS strings

$$Z^{string}(X;p,q,y) = \chi_{orb}(S_p X; q,y) = \prod_{n>0,\, m\geq 0,\, \ell} (1 - p^n q^m y^\ell)^{-c(nm,\ell)} \qquad (5.23)$$

with the coefficients $c(m,\ell)$ determined by the elliptic genus of X,

$$\chi(X;q,y) = \sum_{m,\ell} c(m,\ell) q^m y^\ell. \qquad (5.24)$$

Note that since the strings carry only left-moving excitations, $\overline{L}_0 = 0$, the space-time Hamiltonian p^- and winding number w^- can be identified and thus the partition function represents the *space-time* character

$$Z^{string}(X;p,q,y) = \text{Tr}_{\mathcal{F}}(-1)^F p^{P^+} q^{W^-} y^F. \tag{5.25}$$

This is precisely the object we promised in our discussion to study.

We will parametrize p,q,y as

$$p = e^{2\pi i\sigma},\ q = e^{2\pi i\tau},\ y^{2\pi iz}, \tag{5.26}$$

or equivalently by a 2×2 period matrix

$$\Omega = \begin{pmatrix} \sigma & z \\ z & \tau \end{pmatrix} \tag{5.27}$$

in the Siegel upper half-space, $\det \text{Im}\,\Omega > 0$. The group $Sp(4,\mathbb{Z}) \cong SO(3,2,\mathbb{Z})$ acts on the matrix Ω by fractional linear transformations, $\Omega \to (A\Omega + B)(C\Omega + D)^{-1}$.

Now the claim is that the string partition function $\chi_{orb}(X;p,q,y)$ is almost equal to an automorphic form for the group $SO(3,2,\mathbb{Z})$, of the infinite product type as appear in the work of Borcherds [35]. This is just the string theory generalization of the fact that the Euler number $\chi_{orb}(S_pX)$ is almost a modular form of $SL(2,\mathbb{Z})$. In fact, the Euler number is obtained from the elliptic genus in the limit $y \to 1$, $z \to 0$, where the q-dependence disappears. In this case, the automorphic group degenerates as

$$Sp(4,\mathbb{Z}) \to SL(2,\mathbb{Z}) \times SL(2,\mathbb{Z}) \tag{5.28}$$

where only the first $SL(2,\mathbb{Z})$ factor acts non-trivially on p.

The precise form of the corrections needed to get a true automorphic function $\Phi(p,q,y)$ for a general Calabi-Yau d-fold X has been worked out in detail in [26]. It is defined by the product

$$\Phi(p,q,y) = p^a q^b y^c \prod_{(n,m,\ell)>0} (1 - p^n q^m y^\ell)^{c(nm,\ell)} \tag{5.29}$$

where the positivity condition means: $n,m \geq 0$ with $\ell > 0$ in the case $n = m = 0$. The 'Weyl vector' (a,b,c) is defined by

$$a = b = \chi(X)/24, \qquad c = \sum_\ell -\frac{|\ell|}{4}c(0,\ell). \tag{5.30}$$

Here the coefficients $c(0,\ell)$ are the partial Euler numbers

$$c(0,r - \frac{d}{2}) = \sum_s (-1)^{s+r} h^{r,s} \tag{5.31}$$

One can then show that Φ is an automorphic form of weight $c(0,0)/2$ for the group $O(3,2,\mathbb{Z})$ for a suitable quadratic form of signature $(3,2)$.

The form Φ follows actually from a standard one-loop string amplitude defined as an integral over the fundamental domain [36, 27]. The integrand consists of the genus one partition function of the string on $X \times T^2$ and has a manifest $SO(3,2,\mathbb{Z})$ T-duality invariance. The $SO(3,2,\mathbb{Z})$ appears in the following way. First of all, as explained in the introduction, strings on T^2 have two quantized momenta and two winding numbers, giving the Narain lattice $\Gamma^{2,2}$. For a transversal Calabi-Yau space, there are also the left-moving and right-moving Fermi numbers F_L, F_R. Since we restrict to right-moving ground states in the elliptic genus, only F_L gives another integer conserved quantum number ℓ. Adding this charge to the Narain lattice enlarges it to $\Gamma^{3,2}$. Moreover, it allows us to extend the moduli σ, τ of the two-torus by another complex parameter z that couples to $F_L = \ell$. Technically, z has an interpretation as a Wilson loop that parametrizes the $U(1)_L$ bundle over T^2. Together, σ, τ, z parametrize the lattice $\Gamma^{3,2}$; they can be considered as a point on the symmetric space

$$SO(3,2)/SO(3) \times SO(2) \cong \mathbb{H}^{2,1}. \tag{5.32}$$

Now the strategy is to compute the string partition function through a one-loop amplitude

$$Z^{string}(X;p,q,y) = \exp F^{string}(X;p,q,y). \tag{5.33}$$

Note that F^{string} is the partition function for maps from the world-sheet elliptic curve, with a modulus that we denote as τ', to the space-time that contains an elliptic curve with modulus τ,

$$T^2_{\tau'} \to T^2_{\sigma,\tau,z} \times X. \tag{5.34}$$

It is easy to confuse the two elliptic curves! One now computes an integral over the fundamental domain of the world-sheet modulus τ' that has the form

$$F^{string} = \frac{1}{2} \int \frac{d^2\tau'}{\tau_2'} \sum_{-\frac{d}{2}+1\leq\epsilon\leq\frac{d}{2}} \sum_{\substack{(p_L,p_R) \in \Gamma^{3,2}_\epsilon \\ n \in 2d\mathbb{Z} - \epsilon^2}} e^{i\pi(\tau'p_L^2 - \overline{\tau}'p_R^2)} c_\epsilon(n) e^{\pi i n\tau'/d} \tag{5.35}$$

where the notation $\Gamma^{3,2}_\epsilon$ indicates that $\ell = \epsilon$ (mod $2d$) and where the coefficients $c_\epsilon(n)$ are defined in terms of the expansion coefficients of the elliptic genus of X as $c(m,\ell) = c_\epsilon(2dm - \ell^2)$, with $\epsilon = \lambda$ (mod $2d$). This integral can be computed using the by-now standard techniques of [40, 36, 27, 28]. The final result of the integration is [26],

$$F^{string}(\Omega,\overline{\Omega}) = -\log\left((\det \mathrm{Im}\,\Omega)^{c(0,0)/2}|\Phi(\Omega)|^2\right) \tag{5.36}$$

Since the integral F is by construction invariant under the T-duality group $O(3,2,\mathbb{Z})$, this determines the automorphic properties of Φ. The factor $\det \text{Im} \, \Omega$ transforms with weight -1, which fixes the weight of the form Φ to be $c(0,0)/2$. This formula should be contrasted with the analogous computation for the zero-modes (4.8), $i.e.$, the field theory limit,

$$F^{QFT}(\tau,\bar\tau) = -\log\left((\text{Im}\,\tau)^{c/2}|\eta^c(\tau)|^2\right), \qquad c = \chi(X). \tag{5.37}$$

In the special case of $K3$ the infinite product $\Phi(\Omega)$ is a well-known automorphic form [41], see also [42, 43]. First of all, the elliptic genus of $K3$ is the unique (up to a scalar) weak Jacobi form of weight 0 and index 1. Realizing $K3$ as a Kummer surface (resolving the orbifold $T^4/\mathbb{Z}_2$) we see that the elliptic genus can be written in terms of genus one theta-functions as

$$\chi(K3;p,q) = 2^3 \sum_{even \; \alpha} \frac{\vartheta_\alpha^2(z;\tau)}{\vartheta_a^2(0,\tau)}. \tag{5.38}$$

If we now identify Ω with the period matrix of a genus two Riemann surface, we can rewrite the automorphic form in terms of genus two theta-functions,

$$\Phi(\Omega) = 2^{-12} \prod_{even \; \alpha} \vartheta[\alpha](\Omega)^2 \tag{5.39}$$

In the work of Gritsenko and Nikulin [41] is is shown that Φ also has an interpretation as the denominator of a generalized Kac-Moody algebra. It is a rather obvious conjecture that this GKM should be given by the algebra of BPS states induced by the string interaction. The full story for $K3$ is quite beautiful and explained in [28]. See [44, 45] for more on the connection with GKM's.

Summarizing we have seen the following:

(i) The (BPS) string theory partition function factorizes in left-moving and right-moving contributions that are holomorphic functions of the moduli p,q,y.

(ii) The holomorphic contributions are automorphic forms of weight $-c(0,0)/2$ of the group $SO(3,2,\mathbb{Z})$ if a particular correction is added. This correction takes the form

$$(pq)^{\chi(X)/24} y^c \prod_{\ell>0}(1-y^\ell)^{c(0,\ell)} \prod_{m>0,\ell}(1-q^m y^\ell)^{c(m,\ell)} \tag{5.40}$$

These three factors have the following interpretation. The first factor is again the regulated zero-point energy, very similar to the field theory result. The second factor is due to the bosonic and fermionic zero-modes. (Recall that the low-energy field theory describes general differential forms on $T^2 \times X$.) The third factor is there to restore the symmetry in p and q. It can only be understood using T-duality.

(iii) The full partition function is invariant because the zero-mode contribution adds a non-holomorphic factor $(\det \mathrm{Im}\,\Omega)^{-c(0,0)/2}$.

(iv) The holomorphic contributions are characters of an infinite-dimensional generalized Kac-Moody algebra, directly related to the creation and annihilation operators of the string Fock space and their interactions.

(v) The modularity of the characters, *i.e.*, the transformation properties under the automorphic group $SO(3,2,\mathbb{Z})$, is 'explained' by the relation to a partition function of a string on a two-torus T^2 with an associated line bundle with moduli τ, σ, z and T-duality group $SO(3,2,\mathbb{Z})$.

6   Matrix Strings and Interactions

Up to now we have only considered free theories and observed how in light-cone quantization these models could be reformulated using first-quantized theories on symmetric products. Now we want to take advantage from this relation to include interactions. This has proven possible for two important examples: 1) the ten-dimensional IIA superstring and some of its compactifications, and 2) the class of (2,0) supersymmetric six-dimensional non-abelian string theories. By taking the low-energy limit, similar formulations for the field theory limits follow. The essential starting point in these constructions is the beautiful Ansatz for a non-perturbative formulation of M-theory known as matrix theory [5]. See for example the reviews [8, 7, 9] for more information about matrix theory.

6.1   Supersymmetric Yang-Mills theory

Matrix string theory gives a very simple Ansatz of what non-perturbative IIA string theory looks like in light-cone gauge [12, 13, 11]. It is simply given by the maximally supersymmetric two-dimensional Yang-Mills theory with gauge group $U(N)$ in the limit $N \to \infty$ (or with finite N in DLCQ).

To be more precise, let us consider two-dimensional $U(N)$ SYM theory with 16 supercharges. It can be obtained by dimensionally reducing the $\mathcal{N} = 1$ SYM theory in 10 dimensions. Its field content consists of the following fields. First we pick a (necessarily trivial) $U(N)$ principal bundle P on the world-sheet $S^1 \times \mathbb{R}$. Let A be a connection on this bundle. We further have 8 scalar fields X^i in the vector representation V of $Spin(8)$, and 8 left-moving fermions θ^a in the spinor representation S^+ and 8 right-moving fermions $\overline{\theta}^{\dot{a}}$ in the conjugated spinor representation S^-. All these fields are Hermitean $N \times N$ matrices, or if one wishes sections of the adjoint bundle $\mathrm{ad}(P)$.

The action for the SYM theory reads

$$S_{SYM} = \int d^2\sigma \, \text{Tr} \left(\tfrac{1}{2}|DX^i|^2 + \theta^a \overline{D}\theta^a + \overline{\theta}^{\dot{a}} D \overline{\theta}^{\dot{a}} \right.$$

$$\left. + \frac{1}{2g^2}|F_A|^2 + g^2 \sum_{i<j}[X^i, X^j]^2 + g\,\overline{\theta}^{\dot{a}} \gamma^i_{a\dot{a}}[X^i, \theta^a] \right) \qquad (6.1)$$

Here g is the SYM coupling constant — a dimensionful quantity with dimension 1/length in two dimensions. This means in particular that the SYM model is not conformal invariant. In fact, at large length scales (in the infrared or IR) the model becomes strongly interacting. So we have a one-parameter family of QFT's labeled by the coupling constant g or equivalently a length scale $\ell = 1/g$.

The relation with string theory is the following. First of all for finite N the Hilbert space of states of the SYM theory should be identified with the DLCQ *second-quantized* IIA string Hilbert space. The integer N that gives the rank of the gauge group is then related to the total longitudinal momentum in the usual way as

$$p^+ = N/R, \qquad (6.2)$$

whereas the total light-cone energy is given by

$$p^- = \frac{N}{p^+} H_{SYM} \qquad (6.3)$$

with H_{SYM} the Hamiltonian of the SYM model. Note that in the decompactification of the null circle where we will take $N, R \to \infty$, keeping their ratio finite, only SYM states with energy

$$H_{SYM} \sim \frac{1}{N} \qquad (6.4)$$

will contribute a finite amount to p^-. Finally, the IIA string coupling constant g_s (a dimensionless constant) is identified as

$$g_s = (g\ell_s)^{-1} \qquad (6.5)$$

with ℓ_s the string length, $\alpha' = \ell_s^2$.

From this identification we see that free string theory ($g_s = 0$) is recovered at strong SYM coupling ($g = \infty$). This is equivalent to the statement that free string theory is obtained in the IR limit. In this scaling limit—the fixed point of the renormalization group flow—we expect on general grounds to recover a superconformal field theory with 16 supercharges. We will now argue that this SCFT is the supersymmetric sigma model with target space $S^N \mathbb{R}^8$. We can then use our previous analysis of orbifold sigma models to conclude that the point $g_s = 0$ indeed describes the second-quantized free IIA string.

The analysis proceeds in two steps. First we observe that because of the last two terms in the action (6.1), in the limit $g_s = 0$ which is equivalent to $g = \infty$, the fields X and θ necessarily have to commute. This means that we can write the matrix coordinates as

$$X^i(\sigma) = U(\sigma) \cdot x^i(\sigma) \cdot U^{-1}(\sigma), \qquad (6.6)$$

with $U \in U(N)$ and x^i a diagonal matrix with eigenvalues $x_1^i, \dots, x_N^i$. Now the matrix valued fields $X^i(\sigma)$ are single-valued, being sections of the trivial $U(N)$ vector bundle $\mathrm{ad}(P)$. But this does not imply that the fields $U(\sigma)$ and $x^i(\sigma)$ are too. In fact, it is possible that after a shift $\sigma \to \sigma + 2\pi$ the individual eigenvalues are permuted due to a spectral flow. Only the set of eigenvalues (or more properly the set of common eigenstates) of the commuting matrices X^i is a gauge invariant quantity. So we should allow for configurations of the form

$$x^i(\sigma + 2\pi) = g \cdot x^i(\sigma) \cdot g^{-1}, \qquad (6.7)$$

with $g \in S_N$ the Weyl group of $U(N)$. Effectively this tells us that we are dealing with an orbifold with target space

$$\mathbb{R}^{8N}/S_N = S^N \mathbb{R}^8, \qquad (6.8)$$

given Lie-theoretically as t^8/W with t the Cartan Lie algebra and W the Weyl group of $U(N)$.

As we have analyzed before this implies that the Hilbert space decomposes in superselection sectors labeled by the conjugacy classes $[g]$ of S_N, which in turn are given by partitions of N. This structure indicates that the Hilbert space is a Fock space of second-quantized IIA strings. A sector twisted by

$$g = (n_1) \dots (n_k) \qquad (6.9)$$

describes k strings of longitudinal momentum

$$p_i^+ = \frac{n_i}{R} = \frac{n_i}{N} p_{tot}^+, \qquad i = 1, \dots, k. \qquad (6.10)$$

We have also seen how for a string with a twist (n) of 'length' n the $\mathbb{Z}_n$ projection of the orbifold projects the Hilbert space to a subsector conditioned to

$$L_0 - \overline{L}_0 = 0 \pmod{n} \qquad (6.11)$$

that we now interpret as the usual DLCQ level-matching condition. In the large N limit, also the individual n_i go to infinity, effectively decompactifying the null circle.

The second step consists of analyzing the behaviour of the gauge field. The possibly twisted configurations of $X^i(\sigma)$ break the gauge group $U(N)$ to an abelian subgroup T that commutes with the configuration $X^i(\sigma)$. In fact, if the twist sector is labeled by a partition

$$n_1 + \dots + n_k = N \qquad (6.12)$$

describing k strings of length $n_1, \dots, n_k$, the unbroken gauge group is

$$T \cong U(1)^k. \tag{6.13}$$

Because of the Higgs effect all the broken components of the gauge field acquire masses of the order g and thus decouple in the IR limit. This leaves us with a free abelian gauge theory on $\mathbb{R} \times S^1$. This model has been analyzed in great detail. Dividing by the gauge symmetries leaves us with the holonomy along the S^1

$$Hol(A) = e^{\oint_{S^1} A} \in T \tag{6.14}$$

as the only physical degree of freedom. The gauge theory is therefore described by the quantum mechanics on the torus T with Hamiltonian given by

$$H = -g^2 \Delta \tag{6.15}$$

with Δ the Laplacian on T. The eigenstates are given by the characters of the irreducible representations of T with eigenvalues (energies) g^2 times the second Casimir invariant of the representation. Clearly in the limit $g \to \infty$ only the vacuum state or trivial representation survives. This state has a constant wave function on T which has the interpretation that the abelian gauge field is free to fluctuate, a result from the fact that in strong coupling the action $S = \frac{1}{g^2} \int F^2$ goes to zero. So all-in-all the gauge field sector only contributes a single vacuum state. This completes our heuristic derivation of the IR limit of SYM.

Since two-dimensional gauge theories are so well-behaved it would be interesting to make the above in a completely rigorous statement about the IR fixed point of SYM. One of the points of concern could be complications that emerge if some of the eigenvalues coincide. In that case unbroken non-abelian symmetries appear. As we will show in the next section however, from the SCFT perspective such effects are always irrelevant and thus disappear in the IR limit. In fact, these effects are exactly responsible for the perturbative interactions at finite g.

6.2   Interactions

If the matrix string theory conjecture is correct, for finite coupling constant the SYM theory should reproduce the interacting string. A non-trivial check of this conjecture is to identify the correction for small g_s. This should be given by the joining and splitting interaction of the strings, producing surfaces with nontrivial topology.

This computation was done in [11] where the leading correction was computed. Let us try to summarize this computation. (It is also reviewed in [9].) The idea is to analyze the behaviour of the SYM theory in the neighbourhood of the IR fixed point. In leading order, a deformation to finite g, this is given by the least irrelevant operator in the orbifold CFT. That is, we look for the operator $\mathcal{O}$ in the sigma model that preserves all the supersymmetries and the $Spin(8)$ R-symmetry and that has the smallest scaling dimensions. The deformed QFT then has an action of the form

$$S = S_{SCFT} + (g_s)^{h-2} \int \mathcal{O} + \dots \tag{6.16}$$

with h the total scaling dimension of $\mathcal{O}$. We would like to see that the power of g_s is one (so that $h = 3$) and that this deformation induces the usual joining and splitting interaction.

Note that the Hilbert space of the matrix string was defined with Ramond boundary conditions for the supercurrent $G^{\dot{a}} = \gamma_i^{a\dot{a}} \theta^a \partial x^i$. That is, we have

$$G^{\dot{a}}(\sigma + 2\pi) = G^{\dot{a}}(\sigma). \tag{6.17}$$

We have seen that the ground state space $\mathcal{V}^{(n)}$ of a $\mathbb{Z}_n$ twisted sector $\mathcal{H}^{(n)}$ is isomorphic to the ground state space of a single string

$$\mathcal{V}^{(n)} \cong (V \oplus S^-) \otimes (V \oplus S^+). \tag{6.18}$$

Only the conformal dimensions are rescaled and given by

$$L_0 = \overline{L}_0 = nd/8, \tag{6.19}$$

since the central charge of the SCFT is n times as big. Here d is the complex dimension of the target space, so in our case $d = 4$.

One way to understand this vacuum degeneracy is that the $\mathbb{Z}_n$ action on the n fermions $\theta_1, \dots, \theta_n$ can be diagonalized with eigenvalues $e^{2\pi i k/n}$, $k = 0, \dots, n-1$. That is, there are linear combinations of the θ_k, let us denote them by $\tilde{\theta}_k$ that have boundary conditions

$$\tilde{\theta}_k(\sigma + 2\pi) = e^{\frac{2\pi i k}{n}} \tilde{\theta}_k(\sigma). \tag{6.20}$$

So the linear sum

$$\tilde{\theta}_0 = \theta_1 + \dots + \theta_n \tag{6.21}$$

is always periodic and its zero modes give the 16 fold vacuum degeneracy. A similar story holds for the right-moving fermions.

Since we want to keep Ramond boundary conditions in the interacting theory, the local operator $\mathcal{O}$ that describes the first-order deformation should be in the so-called Neveu-Schwarz (NS) sector. This just tells us that the operator product expansion

$$G^{\dot{a}}(z)\mathcal{O}(w) \tag{6.22}$$

is single-valued in $z - w$. So, using the familiar operator-state correspondence of CFT we have to look in the NS-sector of the Hilbert space. This is of course again labeled by twist fields. The only difference is that the fermions now have an extra minus sign in their monodromy, and satisfy the boundary conditions

$$\tilde{\theta}_k(\sigma + 2\pi) = -e^{\frac{2\pi i k}{n}}\tilde{\theta}_k. \tag{6.23}$$

Now depending on whether n is even or odd there is a periodic fermion or not. So we expect to find only a degeneracy for even n. It is not difficult to compute the conformal dimension of the NS ground state in a $\mathbb{Z}_n$ twisted sector. First of all, both for the bosons and fermions the $\mathbb{Z}_n$ action can be diagonalized. The bosonic twist field that implements a twist with eigenvalue $e^{2\pi i k/n}$ has conformal dimensions $dk(n-k)/2n^2$, with d the complex dimension of the transversal space ($d = 4$ for the IIA string). For the corresponding fermionic twist field we find conformal dimension $dm^2/2n^2$, where $m = \min(k, N-k)$. Adding up all the possible eigenvalues we obtain total conformal dimension

$$h = \begin{cases} n, & n \text{ even,} \\ n - \frac{1}{n}, & n \text{ odd.} \end{cases} \tag{6.24}$$

In particular the lowest dimension $h = 2$ is given by the $\mathbb{Z}_2$ twist field σ. Since $n = 2$ is even, this ground state has the usual degeneracy

$$\sigma \in (V \oplus S^-) \otimes (V \oplus S^+) \tag{6.25}$$

Note that the zero-modes of the superpartner of the twisted boson x^i give this degeneracy. However, the NS ground state is neither supersymmetric nor $Spin(8)$ invariant, and is therefore not a suitable candidate for our operator $\mathcal{O}$.

There is however a small modification that does respect the supersymmetry algebra. In the $\mathbb{Z}_2$ twisted sector the coordinate x^i has a mode expansion

$$\partial x^i = \sum_{n \in \mathbb{Z} + \frac{1}{2}} \alpha_n^i z^{-n-1}. \tag{6.26}$$

We now consider the first excited state

$$\mathcal{O} = \alpha_{-1/2}^i \overline{\alpha}_{-1/2}^j \sigma^{ij} \tag{6.27}$$

of conformal weights $2 + 1 = 3$. (Here σ^{ij} indicates the components of σ in $V \otimes V$.) This operator can be written as

$$\mathcal{O} = G_{-1/2}^{\dot{a}} \overline{G}_{-1/2}^{\dot{b}} \sigma^{\dot{a}\dot{b}} \tag{6.28}$$

and therefore satisfies

$$[G_{-1/2}^{\dot{a}}, \mathcal{O}] = \partial \overline{G}_{-1/2}^{\dot{b}} \sigma^{\dot{a}\dot{b}} \tag{6.29}$$

which is sufficient. Since $\mathcal{O}$ is both supersymmetric and $Spin(*)$ invariant, it is the least irrelevant operator that we were looking for.

What is the interpretation of the field $\mathcal{O}$ in string perturbation theory? It clearly maps superselection sectors with two strings into sectors with one string and vice versa. It is therefore exactly the usual joining and splitting interaction. In fact, the perturbation in the operator $\mathcal{O}$ reproduces the standard light-cone perturbation theory.

There is also a clear geometric interpretation of the twist field interaction $\mathcal{O}$. Consider the manifold $\mathbb{R}^8/\mathbb{Z}_2$ or if one wishes the compact version $T^8/\mathbb{Z}_2$. This is a Calabi-Yau orbifold and defines a perfectly well-behaved superconformal sigma model. One could now try to blow-up the $\mathbb{Z}_2$ singularity to obtain a smooth Calabi-Yau space. It is well-known that this cannot be done without destroying the Calabi-Yau property; the orbifold $\mathbb{R}^8/\mathbb{Z}_2$ is rigid. In the SCFT language this is expressed by the fact that the corresponding deformation does not respect the superconformal algebra. Algebraically, preserving the conformal invariance implies that the operator is marginal with scaling dimension 2. The fact that we found weight 3 is therefore in accordance with the fact that the two-dimensional field theory deforms to a massive field theory with a length scale — two-dimensional SYM.

However. we see that if the transverse target space would have been four-dimensional, the twist field interaction would have $L_0 = \overline{L}_0 = 1$ and would have represented a marginal operator. This is a simple reflection of the fact that the orbifold $\mathbb{R}^4/\mathbb{Z}_2$ or $T^4/\mathbb{Z}_2$ can be resolved to a smooth Calabi-Yau manifold, respectively an hyperkähler asymptotically locally Euclidean (ALE) space or a K3 surface. We therefore turn now to the case where this four-dimensional example becomes relevant.

7 String Theories in Six Dimensions

For superstrings the critical dimension is ten, or transversal dimension eight. However, in the past year there has been growing evidence that there is also a fascinating class of string theories with critical dimension six, with a four-dimensional transversal space. In fact, there is believed to be such a string for every simply-laced Lie group of type A,D,E. We will mainly focus on the $U(k)$ case.

There is very little known about these theories [46, 48, 47]. They have (2,0) supersymmetry with a $Spin(4)$ R-symmetry, do not contain a graviton, and give non-trivial six-dimensional SCFT's in the IR, where the R-symmetry is enlarged to $Spin(5) = Sp(2)$. Roughly their massless modes should be a theory of non-abelian two-form gauge fields, whose three-form field strength is self-dual. For the $U(1)$ theory this can be made precise. The massless modes form the irreducible (2,0) tensor multiplet which consists of one two-form B together with five scalar fields X^I.

Furthermore the string coupling of these microstrings or little strings is believed to be fixed to one (basically because of the self-duality). Since the string coupling cannot be tuned to zero, there is no reason why a free string spectrum should emerge. This is good, because we know that the six-dimensional Green-Schwarz superstring is not Lorentz invariant. It is true however that this string does reproduce the tensor multiplet as its massless sector.

7.1   DLCQ formulations and matrix models

Up to now we only know how to describe these (2,0) strings in a matrix theory DLCQ formulation [14, 49, 50, 51]. We choose a six-dimensional space-time of the form

$$M^{1,5} = (\mathbb{R} \times S^1)^{1,1} \times X^4 \qquad (7.1)$$

with X a (Ricci flat) Riemannian four-manifold. We will often choose X to be compact, which restricts us to either T^4 or $K3$. We fix the longitudinal momentum to be $p^+ = N/R$, with R the radius of the null-circle. The claim is now that this string theory can be described in terms of a two-dimensional sigma model with target space the moduli space

$$\mathcal{M}_{k,N}(X)$$

of $U(k)$ instantons (self-dual connections) on X with total instanton charge $ch_2 = N$. This moduli space is a hyperkähler manifold of real dimension $4Nk$. It has singularities, corresponding to (colliding) point-like instantons. There is however a particularly nice compactification by considering the moduli space

$$\overline{\mathcal{M}}_{k,N}(X)$$

of (equivalence classes) of coherent torsion free sheaves of rank k and $ch_2 = N$. In particular for the case $k = 1$ we find in this way the Hilbert scheme of dimension zero subschemes of length N

$$\overline{\mathcal{M}}_{1,N}(X) = \mathrm{Hilb}^N(X). \qquad (7.2)$$

This space is a intricate smooth resolution of the symmetric space $S^N X$ [52]. The fibers of the projection $\mathrm{Hilb}^N(X) \to S^N X$ over the various diagonals keep track of the particular way the points approach each other. Quite generally, if X is a smooth Calabi-Yau space of complex dimension d then the symmetric product $S^N X$ is also Calabi-Yau manifold, albeit an orbifold, now of dimension Nd. Only for (complex) dimension two, $i.e.$, if X a four-torus or $K3$ surface, is it possible to resolve the singularities of $S^N X$ to produce smooth Calabi-Yau. The Hilbert scheme $\mathrm{Hilb}^N(X)$ provides a canonical construction. For CY d-folds with $d \geq 3$ the Hilbert scheme is not smooth.

We should mention here that all of the spaces $\mathcal{M}_{k,N}$ are to be hyperkähler deformations of $S^{Nk} X$. In particular this implies that their cohomology is given by that of the symmetric product. For more on this issue see [53].

7.2   Deformations and interactions

It is known that the deformation space of any Calabi-Yau space X is locally given by $H^1(T_X) \cong H^{1,d-1}(X)^*$. By a well-known result of Tian and Todorov there are no obstructions to such deformations, and therefore the dimension of the moduli space $\mathcal{M}_X$ of inequivalent complex structures is given by $h^{1,d-1}(X)$. It is not difficult to compute the dimension of the deformation space of the symmetric product $S^N X$ using the above formalism. We see that there is always a contribution given by $H^1(T_X)$. This corresponds to simply deforming the underlying manifold X. However, for dimension $d = 2$ and only for this dimension, there is a second contribution coming from $H^0(X^{(2)})$. In fact, for $d = 2$ we have

$$\dim \mathcal{M}_{S^N X} = \dim \mathcal{M}_X + 1. \tag{7.3}$$

There is a direct geometric interpretation of this extra deformation. $X^{(2)}$ represents the big diagonal in X^N where two points coincide. In the orbifold cohomology of $S^N X$ it contributes the cohomology of X, shifted however in bi-degree by $(d-1, d-1) = (1,1)$. The corresponding deformation corresponds to blowing up in the given complex structure this big diagonal.

The corresponding operator in the SCFT is exactly the same $\mathbb{Z}_2$ twist field that we have discussed before for the type II string. Therefore this deformation can be given an interpretation as tuning the string coupling constant [53].

Acknowledgements

A very much shortened version of these notes can be found in [54]. I wish to thank the organizers of the *Geometry and Duality Workshop* at the Institute for Theoretical Physics, UC Santa Barbara, January 1998 and the *Spring School on String Theory and Mathematics,* Harvard University, May 1998, where the material of these lectures was presented.

Bibliography

[1] B. Zwiebach, *Closed string field theory: quantum action and the B-V master equation,* Nucl. Phys. **B390** (1993) 33–152, hep-th/9206084.

[2] C. Hull and P. Townsend, *Unity of superstring dualities,* Nucl. Phys. **B 438** (1995) 109, hep-th/9410167.

[3] J. Polchinski, *Dirichlet-branes and Ramond-Ramond charges,* Phys. Rev. Lett. **75** (1995) 4724–4727, hep-th/9510017.

[4] E. Witten, *String theory in various dimensions,* Nucl. Phys. **B 443** (1995) 85, hep-th/9503124.

[5] T. Banks, W. Fischler, S. H. Shenker, and L. Susskind, *M Theory as a matrix model: a conjecture,* Phys. Rev. **D55** (1997) 5112–5128, hep-th/9610043.

[6] A. Bilal, *M(atrix) theory : a pedagogical introduction*, hep-th/9710136.

[7] T. Banks, *Matrix theory*, hep-th/9710231.

[8] D. Bigatti and L. Susskind, *Review of matrix theory*, hep-th/9712072.

[9] R. Dijkgraaf, E. Verlinde, and H. Verlinde, *Notes on matrix and micro strings*, hep-th/9709107.

[10] R. Dijkgraaf, G. Moore, E. Verlinde, and H. Verlinde, *Elliptic genera of symmetric products and second quantized strings*, Commun. Math. Phys. **185** (1997) 197–209, hep-th/9608096.

[11] R. Dijkgraaf, E. Verlinde, and H. Verlinde, *Matrix string theory*, Nucl. Phys. **B500** (1997) 43–61, hep-th/9703030.

[12] L. Motl, *Proposals on non-perturbative superstring interactions*, hep-th/9701025.

[13] T. Banks and N. Seiberg, *Strings from matrices*, Nucl. Phys. **B497** (1997) 41–55, hep-th/9702187.

[14] R. Dijkgraaf, E. Verlinde, and H. Verlinde, *5D Black holes and matrix strings*, Nucl. Phys. **B506** (1997) 121–142, hep-th/9704018.

[15] E. Witten, *Geometry and physics*, ICM, Berkeley (1988).

[16] L. Göttsche, *The Betti numbers of the Hilbert scheme of points on a smooth projective surface*, Math. Ann. **286** (1990) 193–207; *Hilbert Schemes of Zero-dimensional Subschemes of Smooth Varieties*, Lecture Notes in Mathematics **1572**, Springer-Verlag, 1994.

[17] L. Göttsche and W. Soergel, *Perverse sheaves and the cohomology of Hilbert schemes of smooth algebraic surfaces*, Math. Ann. **296** (1993) 235–245.

[18] J. Cheah, *On the cohomology of Hilbert schemes of points*, J. Alg. Geom. **5** (1996) 479–511.

[19] F. Hirzebruch and T. Höfer, *On the Euler Number of an orbifold*, Math. Ann. **286** (1990) 255.

[20] C. Vafa and E. Witten, *A strong coupling test of S-duality*, Nucl. Phys. **B431** (1994) 3–77, hep-th/9408074.

[21] G. Segal, *Equivariant K-theory and symmetric products*, manuscript and lecture at Aspen Center of Physics, August 1996.

[22] I.G. Macdonald, *The Poincaré polynomial of a symmetric product*, Proc. Camb. Phil. Soc **58** (1962) 563–568.

[23] A. Schwimmer and N. Seiberg, *Comments on the N=2, N=3, N=4 superconformal algebras in two-dimensions*, Phys. Lett. **184B** (1987) 191.

[24] P.S. Landweber Ed., *Elliptic Curves and Modular Forms in Algebraic Topology* (Springer-Verlag, 1988).
E. Witten, Commun.Math.Phys. **109** (1987) 525.
A. Schellekens and N. Warner, Phys. Lett, **B177** (1986) 317; Nucl.Phys. **B287** (1987) 317.
O. Alvarez, T.P. Killingback, M. Mangano, and P. Windey, *The Dirac-Ramond operator in string theory and loop space index theorems*. Nucl. Phys. B (Proc. Suppl.) **1A** (1987) 89; *String theory and loop space index theorems*, Commun. Math. Phys.

111 (1987) 1.

T. Eguchi, H. Ooguri, A. Taormina, S.-K. Yang, *Superconformal Algebras and String Compactification on Manifolds with SU(N) Holonomy*, Nucl.Phys. **B315** (1989) 193. T. Kawai, Y. Yamada and S.-K. Yang, *Elliptic Genera and N=2 Superconformal Field Theory*, Nucl. Phys. **B414** (1994) 191-212.

[25] M. Eichler and D. Zagier, *The Theory of Jacobi Forms* (Birkhäuser, 1985).

[26] C.D.D. Neumann, *The elliptic genus of Calabi-Yau 3- and 4-folds, product formulae and generalized Kac-Moody Algebras*, hep-th/9607029.

[27] T. Kawai, *N = 2 Heterotic string threshold correction, K3 surface and generalized Kac-Moody superalgebra*, Phys. Lett. **B372** (1996) 59–64, hep-th/9512046.

[28] T. Kawai, *K3 surfaces, Igusa cusp form and string theory*, hep-th/9710016.

[29] E. Witten, *Mirror manifolds and topological field theory*, in *Essays on Mirror manifolds*, Ed. S-T Yau (International Press, Hong Kong, 1992).

[30] R. Dijkgraaf, E. Verlinde and H. Verlinde, *Counting dyons in N = 4 string theory*, Nucl. Phys. **B484** (1997) 543. hep-th/9607026.

[31] L. Dixon, J. Harvey, C. Vafa, and E. Witten, Nucl. Phys. **B261** (1985) 620; Nucl. Phys. **B274** (1986) 285.

[32] A. Strominger and C. Vafa, *Microscopic origin of the Bekenstein-Hawking entropy*, Phys. Lett. **B379** (1996) 99–104, hep-th/9601029.

[33] J.M. Maldacena and L. Susskind, *D-branes and fat black holes*, Nucl. Phys. **B475** (1996) 679, hep-th/9604042.

[34] R. Dijkgraaf, E. Verlinde and H. Verlinde, *BPS spectrum of the five-brane and black bole entropy*, Nucl. Phys. **B486** (1997) 77–88, hep-th/9603126.

[35] R. E. Borcherds, *Automorphic forms on $O_{s+2,2}(R)$ and Infinite Products*, Invent. Math. **120** (1995) 161.

[36] J. Harvey and G. Moore, *Algebras, BPS states, and strings*, Nucl. Phys. **B463** (1996) 315-368, hep-th/9510182.

[37] N. Seiberg, *Why is the matrix model correct?*, Phys. Rev. Lett. **79** (1997) 3577-3580, hep-th/9710009.

[38] S. Sethi, C. Vafa, and E. Witten, *Constraints on low-dimensional string compactifications*, Nucl. Phys. **B480** (1996) 213-224, hep-th/9606122.

[39] K. Dasgupta and S. Mukhi, *A note on low-dimensional string compactifications*, Phys. Lett. **B398** (1997) 285-290, hep-th/9612188.

[40] L. Dixon, V. Kaplunovsky, and J. Louis, *Moduli-dependence of string loop corrections to gauge coupling constants*, Nucl. Phys. **B307** (1988) 145.

[41] V.A. Gritsenko and V.V. Nikulin, *Siegel Automorphic Form Corrections of Some Lorentzian Kac-Moody Algebras*, Amer. J. Math. **119** (1997), 181–224, alg-geom/9504006; *The Igusa modular forms and "the simplest" Lorentzian Kac-Moody algebras*, alg-geom/9603010.

[42] V.A. Gritsenko and V.V. Nikulin, *Automorphic Forms and Lorentzian Kac–Moody Algebras*, part I and part II, alg-geom/9610022 and alg-geom/9611028.

[43] A.J. Feingold and I.B. Frenkel, *A hyperbolic Kac-Moody algebra and the theory of Siegel modular forms of genus 2*, Math. Ann. **263** (1983) 87–114.

[44] J.A. Harvey and G. Moore, *On the algebra of BPS states*, Commun. Math. Phys. **197** (1998) 489–519, hep-th/9609017.

[45] C.D.D. Neumann, *Perturbative BPS-algebras in superstring theory*, Nucl. Phys. **B499** (1997) 596–620, hep-th/9702197.

[46] E. Witten, *Some comments on string dynamics*, hep-th/9510135.

[47] N. Seiberg, *Notes on theories with 16 supercharges*, hep-th/9705117.

[48] N. Seiberg, *Matrix description of M-theory on T^5 and $T^5/\mathbb{Z}_2$*, Phys. Lett. **B408** (1997) 98–104, hep-th/9705221.

[49] O. Aharony, M. Berkooz, S. Kachru, N. Seiberg, and E. Silverstein, *Matrix description of interacting theories in six dimensions*, Adv. Theor. Math. Phys. **1** (1998) 148–157, hep-th/9707079.

[50] E. Witten, *On the conformal field theory of the Higgs branch*, J. High Energy Phys. **07** (1997) 003, hep-th/9707093.

[51] O. Aharony, M. Berkooz, N. Seiberg, *Light-cone description of (2,0) superconformal theories in six dimensions*, Adv. Theor. Math. Phys. **2** (1998) 119–153, hep-th/9712117.

[52] H. Nakajima, *Heisenberg algebra and Hilbert schemes of points on projective surfaces*, alg-geom/9507012.

[53] R. Dijkgraaf, *Instanton strings and hyperkähler geometry*, hep-th/9810210.

[54] R. Dijkgraaf, *The mathematics of fivebranes*, in *Proceedings of the ICM Berlin 1998*, Doc. Math. III (1998) 133–142, hep-th/9810157.

Addresses

Robbert Dijkgraaf
Korteweg-de Vries Instituut voor Wiskunde
Universiteit van Amsterdam
Plantage Muidergracht 24
NL-1018 TV Amsterdam
The Netherlands
e-mail: rhd@wins.uva.nl

Gerard van der Geer
Korteweg-de Vries Instituut voor Wiskunde
Universiteit van Amsterdam
Plantage Muidergracht 24
NL-1018 TV Amsterdam
The Netherlands
e-mail: geer@wins.uva.nl

Carel Faber
Department of Mathematics
401 Mathematical Sciences
Oklahoma State University
Stillwater, OK 74078-1058
U.S.A.
e-mail: cffaber@math.okstate.edu

Institutionen för Matematik
Kungl Tekniska Högskolan
S-100 44 Stockholm
Sweden
e-mail: carel@math.kth.se

Richard Hain
Department of Mathematics
Duke University
Durham, NC 27708-0320
U.S.A.
e-mail: hain@math.duke.edu

Frans Oort
Mathematisch Instituut
Universiteit Utrecht
P.O. Box 80.010
NL-3508 TA Utrecht
The Netherlands
e-mail: oort@math.uu.nl

Eduard Looijenga
Mathematisch Instituut
Universiteit Utrecht
P.O. Box 80.010
NL-3508 TA Utrecht
The Netherlands
e-mail: looijeng@math.uu.nl